I0064981

# Introduction to Polymer Chemistry

# Introduction to Polymer Chemistry

Walter O'Hara

**NY** RESEARCH
P R E S S
New York

Published by NY Research Press
118-35 Queens Blvd., Suite 400,
Forest Hills, NY 11375, USA
www.nyresearchpress.com

Introduction to Polymer Chemistry
Walter O'Hara

© 2019 NY Research Press

International Standard Book Number: 978-1-63238-698-4 (Hardback)

This book contains information obtained from authentic and highly regarded sources. All chapters are published with permission under the Creative Commons Attribution Share Alike License or equivalent. A wide variety of references are listed. Permissions and sources are indicated; for detailed attributions, please refer to the permissions page. Reasonable efforts have been made to publish reliable data and information, but the authors, editors and publisher cannot assume any responsibility for the validity of all materials or the consequences of their use.

**Trademark Notice:** Registered trademark of products or corporate names are used only for explanation and identification without intent to infringe.

**Cataloging-in-Publication Data**

Introduction to polymer chemistry / Walter O'Hara.
p. cm.
Includes bibliographical references and index.
ISBN 978-1-63238-698-4
1. Polymers. 2. Polymerization. 3. Chemistry. I. O'Hara, Walter.
QD381 .I58 2019
547.28--dc23

# Contents

**Permissions**

**Index**

# Preface

A polymer is a macromolecule that is composed of a number of repeated subunits. It is formed due to the polymerization of monomers. The description of a polymer can be done on the basis of its degree of polymerization, tacticity, molar mass distribution, degree of branching, cystallinity, melting temperature, etc. The branch of chemistry that studies the structure, properties and synthesis of polymers and macromolecules is known as polymer chemistry. The study of synthetic and organic polymeric compositions are primarily studied in this domain. This includes rubber, plastics, fibers and composites. Almost all synthetic polymers are obtained from petrochemicals. This book is a compilation of chapters that discuss the most vital concepts in the field of polymer chemistry. Most of the topics introduced herein cover new techniques and applications of this discipline. It is a complete source of knowledge on the present status of this important field.

To facilitate a deeper understanding of the contents of this book a short introduction of every chapter is written below:

Chapter 1- A large molecule or macromolecule which is composed of repeated subunits is known as polymer. This is an introductory chapter which will discuss briefly all the different types of polymers such as natural polymer, synthetic polymers, biopolymers and bioplastics, among many others.

Chapter 2- The most basic characeteristics of a polymer are the identity of its monomer and its microstructure. Such properties influence the bulk physical properties of the polymer. This chapter delves into the characteristics and properties of polymers such as melting and boiling point, tensile strength, viscoelasticity, optical properties, crystallization of polymers, etc.

Chapter 3- The polymer which has a skeletal structure but does not have carbon atoms in its backbone is referred to as inorganic polymer. The chapter closely examines some inorganic polymers such as polyphosphazene, ionomer, geopolymer and conductive polymers to provide an extensive understanding of the subject.

Chapter 4- A fluorocarbon-based polymer which has multiple carbon-fluorine bonds is termed as a fluoropolymer. It has a high resistance to acids, bases and solvents. Polytetrafluoroethylene, ECTFE perfluoroalkoxy alkanes, polyvinylidene fluoride, etc. are some of the important fluoropolymers covered in this chapter.

Chapter 5- The sub-field of materials science that deals with polymers is known as polymer science. Polymer architecture studies the way branching leads to a deviation from a linear polymer chain. The aim of this chapter is to explore the important aspects of polymer science and architecture such as branched polymers, two-dimensional polymers, etc. for a comprehensive understanding of the subject matter.

Chapter 6- Polymerization is the process of forming polymer chains or three-dimensional networks by reacting monomer molecules together in a chemical reaction. All the different types of polymerization such as radical polymerization, step-growth polymerization and chain-growth polymerization, etc. have been carefully analyzed in this chapter.

Finally, I would like to thank the entire team involved in the inception of this book for their valuable time and contribution. This book would not have been possible without their efforts. I would also like to thank my friends and family for their constant support.

**Walter O'Hara**

# Polymer and its Types

A large molecule or macromolecule which is composed of repeated subunits is known as polymer. This is an introductory chapter which will discuss briefly all the different types of polymers such as natural polymer, synthetic polymers, biopolymers and bioplastics, among many others.

Polymer is a class of natural or synthetic substances composed of very large molecules, called macromolecules that are multiples of simpler chemical units called monomers. Polymers make up many of the materials in living organisms, including, for example, proteins, cellulose, and nucleic acids. Moreover, they constitute the basis of such minerals as diamond, quartz, and feldspar and such man-made materials as concrete, glass, paper, plastics, and rubbers.

The word *polymer* designates an unspecified number of monomer units. When the number of monomers is very large, the compound is sometimes called a high polymer. Polymers are not restricted to monomers of the same chemical composition or molecular weight and structure. Some natural polymers are composed of one kind of monomer. Most natural and synthetic polymers, however, are made up of two or more different types of monomers; such polymers are known as copolymers.

Natural rubber Latex tapped from a rubber tree (*Hevea brasiliensis*) in Malaysia

Organic polymers play a crucial role in living things, providing basic structural materials and participating in vital life processes. For example, the solid parts of all plants are made up of polymers. These include cellulose, lignin, and various resins. Cellulose is a polysaccharide, a polymer that is composed of sugar molecules. Lignin consists of a complicated three-dimensional network of polymers. Wood resins are polymers of a simple hydrocarbon, isoprene. Another familiar isoprene polymer is rubber.

Other important natural polymers include the proteins, which are polymers of amino acids, and the nucleic acids, which are polymers of nucleotides—complex molecules composed of nitrogen-containing bases, sugars, and phosphoric acid. The nucleic acids carry genetic information in the cell. Starches, important sources of food energy derived from plants, are natural polymers composed of glucose.

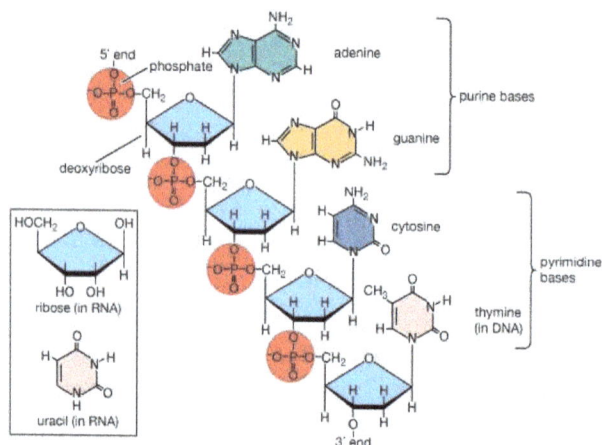

Portion of polynucleotide chain of deoxyribonucleic acid (DNA). The inset shows the corresponding pentose sugar and pyrimidine base in ribonucleic acid (RNA)

Many inorganic polymers also are found in nature, including diamond and graphite. Both are composed of carbon. In diamond, carbon atoms are linked in a three-dimensional network that gives the material its hardness. In graphite, used as a lubricant and in pencil "leads," the carbon atoms link in planes that can slide across one another.

Synthetic polymers are produced in different types of reactions. Many simple hydrocarbons, such as ethylene and propylene, can be transformed into polymers by adding one monomer after another to the growing chain. Polyethylene, composed of repeating ethylene monomers, is an addition polymer. It may have as many as 10,000 monomers joined in long coiled chains. Polyethylene is crystalline, translucent, and thermoplastic—i.e., it softens when heated. It is used for coatings, packaging, molded parts, and the manufacture of bottles and containers. Polypropylene is also crystalline and thermoplastic but is harder than polyethylene. Its molecules may consist of from 50,000 to 200,000 monomers. This compound is used in the textile industry and to make molded objects.

Other addition polymers include polybutadiene, polyisoprene, and polychloroprene, which are all important in the manufacture of synthetic rubbers. Some polymers, such as polystyrene, are glassy and transparent at room temperature, as well as being thermoplastic. Polystyrene can be colored any shade and is used in the manufacture of toys and other plastic objects.

Polystyrene packaging.

If one hydrogen atom in ethylene is replaced by a chlorine atom, vinyl chloride is produced. This polymerizes to polyvinyl chloride (PVC), a colorless, hard, tough, thermoplastic material that can be manufactured in a number of forms, including foams, films, and fibers. Vinyl acetate, produced by the reaction of ethylene and acetic acid, polymerizes to amorphous, soft resins used as coatings and adhesives. It copolymerizes with vinyl chloride to produce a large family of thermoplastic materials.

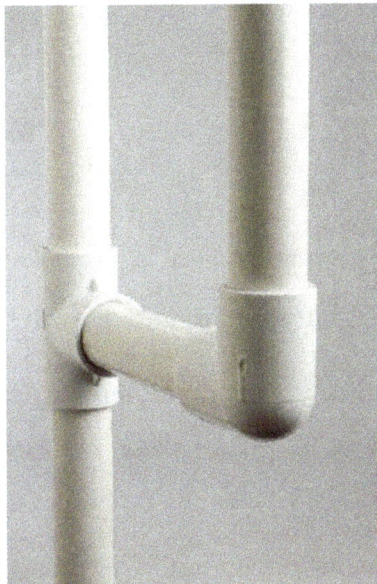

Polyvinyl chloride (PVC) pipes.

Many important polymers have oxygen or nitrogen atoms, along with those of carbon, in the backbone chain. Among such macromolecular materials with oxygen atoms are polyacetals. The simplest polyacetal is polyformaldehyde. It has a high melting point and is crystalline and resistant to abrasion and the action of solvents. Acetal resins are more like metal than are any other plastics and are used in the manufacture of machine parts such as gears and bearings.

A linear polymer characterized by a repetition of ester groups along the backbone chain is called a polyester. Open-chain polyesters are colorless, crystalline, thermoplastic materials. Those with high molecular weights (10,000 to 15,000 molecules) are employed in the manufacture of films, molded objects, and fibers such as Dacron.

The polyamides include the naturally occurring proteins casein, found in milk, and zein, found in corn (maize), from which plastics, fibers, adhesives and coatings are made. Among the synthetic polyamides are the urea-formaldehyde resins, which are thermosetting. They are used to produce molded objects and as adhesives and coatings for textiles and paper. Also important are the polyamide resins known as nylons. They are strong, resistant to heat and abrasion, noncombustible, and nontoxic, and they can be coloured. Their best-known use is as textile fibres, but they have many other applications.

**Formation of nylon**

hexamethylenediamine

$H_2N-CH_2CH_2CH_2CH_2CH_2CH_2-NH_2$

adipic acid

$HO_2CCH_2CH_2CH_2CH_2CO_2H$

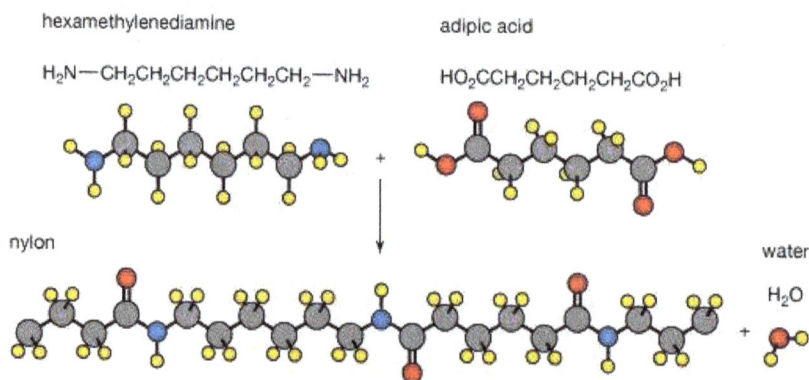

nylon

water

$H_2O$

The formation of nylon, a polymer

Another important family of synthetic organic polymers is formed of linear repetitions of the urethane group. Polyurethanes are employed in making elastomeric fibres known as spandex and in the production of coating bases and soft and rigid foams.

A different class of polymers are the mixed organic-inorganic compounds. The most important representatives of this polymer family are the silicones. Their backbone consists of alternating silicon and oxygen atoms with organic groups attached to each of the silicon atoms. Silicones with low molecular weight are oils and greases. Higher-molecular-weight species are versatile elastic materials that remain soft and rubbery at very low temperatures. They are also relatively stable at high temperatures.

Silicone caulk being dispensed from a caulking gun.

## Applications of Polymers

Agriculture and Agribusiness

- Polymeric materials are used in and on soil to improve aeration, provide mulch, and promote plant growth and health.

Medicine

- Many biomaterials, especially heart valve replacements and blood vessels, are made of polymers like Dacron, Teflon and polyurethane.

Consumer Science

- Plastic containers of all shapes and sizes are light weight and economically less expensive than the more traditional containers. Clothing, floor coverings, garbage disposal bags, and packaging are other polymer applications.

Industry

- Automobile parts, windshields for fighter planes, pipes, tanks, packing materials, insulation, wood substitutes, adhesives, matrix for composites, and elastomers are all polymer applications used in the industrial market.

Sports

- Playground equipment, various balls, golf clubs, swimming pools, and protective helmets are often produced from polymers.

# Natural Polymers

Natural polymers can be defined as polymers derived from natural sources. Polymers of natural origin are biologically recognizable materials that present all natural moieties necessary for proper cell and protein interactions. However, natural polymers also have some limitations such as limited availability because they are derived from finite natural sources, immunogenicity, and the risk of pathogen transmission.

These polymers are formed either by the process of addition polymerization or condensation polymerization. Polymers are extensively found in nature. Our body too is made up of many natural polymers like nucleic acids, proteins, etc. The Cellulose is another natural polymer, which is a main structural component of the plants. Most of the natural polymers are formed from the condensation polymers and this formation from the monomers; water is obtained as a by-product.

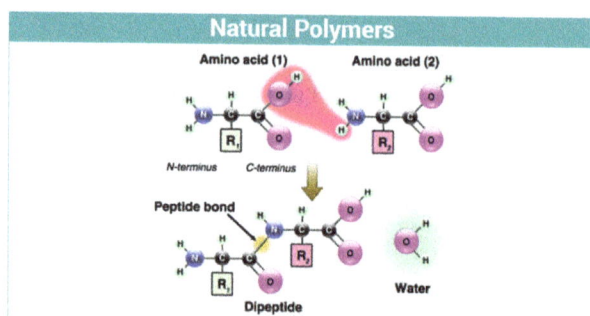

Some of the Natural polymers also include DNA and RNA, these polymers are very much important in all the life processes of all the living organisms. This messenger RNA is the one that makes possible peptides, proteins, and enzymes in a living body. Enzymes inside the living organisms help the reactions to happen and the peptides makes up the structural components of hair, skin, and also the horns of a rhino. The other natural polymers are polysaccharides or called as sugar polymers and polypeptides such as keratin, silk, and the hair. Natural rubber is also a natural polymer, which is made of hydrogen and carbon.

## Examples of Natural Polymers

There are about many examples of natural polymers, which occur in nature. A brief description on some of them are listed below:

- Proteins and Polypeptides- Proteins are the basic type of natural polymers, which constitutes in almost all the living organisms. Proteins are said to be most versatile in nature. They can also be as catalysts. Some of the proteins are called as enzymes. These enzymes are responsible for various chemical reactions occurring in our body and it happens about million times faster even without these enzymes. One type of protein in our blood called as hemoglobin carries the oxygen from lungs to the cells of a human body.

- A protein is usually a naturally occurring type of polyamide. This polymer consists of an amide group present in the backbone chain of human body.

- Collagen– Collagen is one of the natural polymers and is a protein. It makes up the connective tissue present in the skin of human beings. This Collagen-polymer is also a fiber that creates an elastic layer below the skin and thus helps in keeping it supple and smooth.

- Latex- Latex is known to be a kind of rubber, and rubber is a natural polymer. This latex occurs in both the forms either synthetic or natural. The natural form of latex is mainly collected from the rubber trees and it is also found in variety of plants, which includes the milkweed. It can also be prepared artificially by the process of building up long chains of molecules of styrene.

- Cellulose– Cellulose is one of the most abundant organic compounds found on the Earth and moreover the purest form of natural cellulose is the cotton. The paper manufactured from the woods of trees and also the supporting materials in leaves and plants mainly comprise of cellulose. Like the amylose, it is also a polymer, which is made from the monomers of glucose.

- Starch– Starch is the derivative of condensation polymerization and consists of glucose monomers, which further split into water molecules when combined chemically. Starch is also a member of basic food groups called the carbohydrates and it is found in the grains, cereal and potatoes. Starch is a polymer of monosaccharide glucose. The molecules of starch consists of 2 kinds of glucose polymers namely amylopectin and amylose, which are the main component of starch in most of the plants.

## Wool

Wool polymer is a linear, alpha-keratin polymer, which has a helical configuration. Steps in the formation of wool polymer are not known. So the repeating unit of wool polymer is amino acid, which is linked to each other by the peptide bond (-CO-NH-).

As a result, it is not possible to determine the extent or degree of polymerization for wool. It consists of a long polypeptide chain constructed from 18 amino acids:

- Wool polymer is about 140 nm and about 1nm thick.

- In its normal relaxed state, the wool polymer has alpha keratin structure.

- Stretching of the wool fiber tend to stretch, straighten with unfolded configuration called beta-keratin. A beta-keratin wool polymer always tends to return to its relaxed alpha keratin structure.

- Amorphous: Wool polymer system is extremely amorphous, as it is about 25 to 30% crystalline. The spiraling of the proto-fibrils, micro-fibrils and macro-fibrils does not imply a well-aligned polymer system.

## Complex Structure of Wool

The complexity of the wool polymer is due to important chemical groupings it contains and the inter-polymer forces of attraction.

I. Polar peptide groups: The oxygen of the carbonyl groups (-CO-) is slightly negatively charged and as a result will form hydrogen bonds with the slightly positively charged hydrogen of the amino groups (-NH-) of another peptide group.

II. Salt linkages or ionic bonds: carboxyl radicals (-COOH) and (-NH2) as side groups of amino acids which are basically the acidic and basic groups, salt linkages or ionic bond will forms.

III. Covalent bonds: cystine, the sulphur containing amino acid which is present in wool, makes the wool polymer system the only one with cystine linkages, also known as di-sulphide bonds. Cystine bonds are covalent bonds, they occur within and between wool polymers.

## Polymer System of Wool Easily takes Dye Molecules

It is due to the polarity of its polymers and its amorphous nature. The polarity will readily attract any polar dye molecules and draw them into the polymer system. The inter-polymer spaces in the crystalline regions of the polymer system are too small and prevent the relatively large and bulky dye molecules from entering. Therefore the dye molecules can enter the amorphous regions of the polymer system of wool.

## Stress Strain Curve for Wool

Commonly, the stress-strain curve of a single wool fiber which provides tensile characteristics is depicted by 3 different deformation regions, namely, initial Hookean region from 0 to 2% strain corresponding to the reversible deformations of mainly bond angles and lengths, a yield region ranging from 2% to 25-30% in which the α- helices of micro-fibrils unfold and are replaced by β-pleated sheets, and a post yield region beyond 30% in which some irreversible degradation process occur. In one of the interesting studies of mechanical behavior of wool fibers as a function of temperature it was reported that with increasing temperature, tensile properties and durability of the wool fibers decreased considerably. A great decrease on tensile properties was seen at temperatures higher than ~200°C.

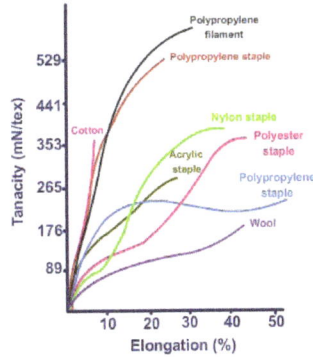

## Physical Properties of Wool

### Tenacity

Wool is a complicated weak polymer. The low tensile strength is because of comparatively fewer hydrogen bonds. When it absorbs moisture, the water molecules steadily force sufficient polymers apart to cause a significant number of hydrogen bonds to break. The water molecules also hydrolyze several salt linkages in the amorphous regions of the strand. Breakage and hydrolysis of these inter-polymer forces of attraction are explicit as swelling of the polymer and result in loss of strength of the wet woolen material.

### Elasticity and Resiliency

This is elastic and resilient. Covalent bonds can stretch, but they are strong. The disulphide bonds in the amorphous parts of the strand or polymer are able to stretch when the strand is extended. When the strand is released the disulphide bonds pull the protein molecules back into their original positions.

If there are too few disulphide linkages as when the strand has been weakened by alkali or if the extension is great enough to break some of the covalent bonds, then some polypeptide chains will slide past one another. This causes a permanent extension of the wool. The natural crispness of the polymer also supports it to regain its real shape.

### Hydroscopic Nature

It has the very absorbent nature because of the polarity of the peptide group, the salt linkages and the amorphous nature of the polymer system. The peptide groups and salt linkages easily attract water molecules, which enter the amorphous polymer system of the polymer. In comparatively dry weather wool may develop static electricity. This is since these are hot enough. Water molecules in the polymer system support to distribute any static electricity, which might develop.

## Density

It has a comparatively low density and therefore polymer is light with regard to their visible weight.

## Conductivity of Heat

It has a low conductivity of heat and therefore makes it ideal for cold weather. The resiliency of the polymer is significant in the warmth properties of the fabric. Wool polymer does not pack well in yarns because of the crimp and scales, and this makes wool fabric process and capable of inserting much air. Air is one of the best insulators since it keeps body heat close to the body. The medulla of the wool polymer comprises air spaces that increase the insulating power of the polymer.

This strand can take up moisture in vapor form. Absorbency is a factor also in the warmth of clothing. In winter, when people go from a dry indoor atmosphere into the damp outdoor air, the heat developed by the polymer in absorbing moisture keeps to protect their bodies from the impact of the cold atmosphere.

## Dimensional Stability

It has poor dimensional stability and therefore shrinks easily. Felting or shrinkage results since under mechanical action, such as agitation, friction and pressure in the presence of heat and moisture, it tends to move root wards, and the edges of the scales interlock prohibiting the polymer from returning to its original position. This results in the fabric becoming thicker and smaller, that is it shrinks or felts.

## Chemical Properties of Wool

### Effect of Acids

Concentrated acids damage it since they hydrolyze the salt linkages and hydrogen bonds. Dilute acids do not affect it.

### Effect of Alkali

It easily dissolves in alkaline solutions. Alkalis hydrolyze the disulphide bonds; hydrogen bonds and salt linkages of wool and cause the polymers to separate from each other, which is looked as dissolution of the polymers. Hydrolysis of the peptide bonds of wool polymers lead to polymer fragmentation and total destruction of the strand. Prolonged exposure to alkalis causes hydrolysis of the peptide bonds of wool polymers lead to polymer fragmentation and total destruction of the polymer.

## Effect of Bleach

Chlorine bleach is ordinary harmful to the wool. KMnO4, Na2O2 are utilized for bleaching.

## Effect of Sunlight and Weather

Effect of sunlight's ultra-violate radiation tends to yellow white or dull colored fabrics. The ultra-violate causes the peptide and disulphide bonds to sever. This leads to polymer degradation products on the surface of the polymer. As a consequence the strand not only absorbs more light but also scatters the incident light to a greater extent. The prolonged exposure to sunlight weakens the polymer very much.

## Colorfastness

Like cotton wool is easy to dye. Acid dyes, chrome and mordant dyes are utilized to dye this. The dye molecules are attracted into the amorphous areas of wool.

## Silk

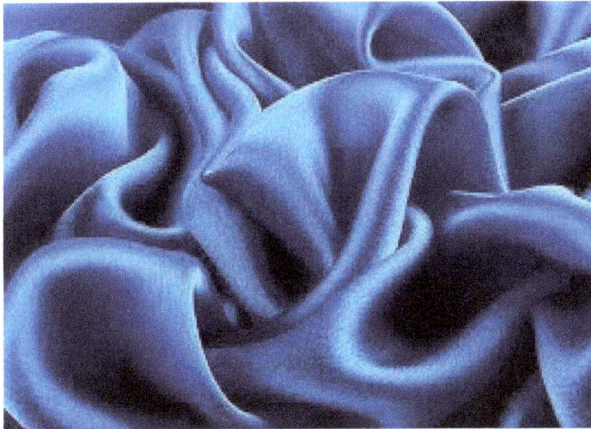

## The Silk Polymer and its System

Silk polymer is linear, fibroin polymer. It is composed of 16 different amino acids. The silk polymer is not composed of any amino acids containing sulphur, so it does not contain any di-sulphid bonds. Silk polymer occurs only in the beta-configuration. Length of the silk polymer is about 140nm, which is slightly longer than wool polymer, and about 0.9 nm thick. The important chemical groupings of the silk polymer are the peptide groups, which give rise to hydrogen bonds, and the carboxyl and amine groups, which give rise to the salt linkages.

Polymer system of silk is composed of layers of folded linear polymers. This results in 65-70% crystalline polymer system. The major forces of attraction between silk polymers are hydrogen bonds, which are effective across a distance of less than 0.5nm. This ensures that fibroin polymers must be closer than this given distance.

## Physical Properties of Silk

## Microscopy of Silk Fiber

Cross section of raw silk is roughly elliptical. Figure shows that triangular twin fibroin filaments, covered by sericin, face each other. The beauty, softness and luster of silk are due to the triangular cross section of the silk filament. After degumming process, twin fibroin filaments separates into individual filaments giving finer and more lustrous fiber. It lacks longitudinal features along the longitudinal-section.

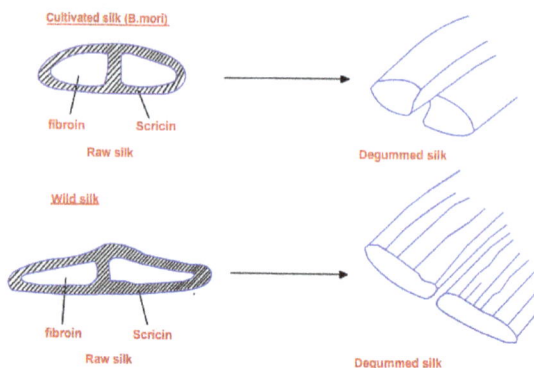

Figure: Cross section of silk fiber

## Appearance

Silk filaments are 600-1700 m long and diameter ranges from 12-30 μm depending upon the health, diet and state under which the silk larvae extruded the silk filaments. So fiber length to breadth ratio is 2000:1. It is off-white to yellow in color.

## Hygroscopic Nature

Moisture regain of silk is about 11% compared to cotton, which has 10.5% this is due to the very crystalline polymer system. The amount of moisture absorbed by silk depends on whether it is raw or degummed silk or on the species of silk and in the humidity.

## Mechanical Properties

Silk is strong fiber. A filament of silk is stronger than an equal diameter of steel filament. This strength is due to linear, beta configuration polymers and very crystalline polymer system. These factors lead to formation of many hydrogen bonds in a much more regular manner.

Its tenacity varies from 2.4 to 5.1 gms per denier. The wet strength of the fiber is about 80-85% of the dry strength. This is because water molecules hydrolyzing a significant number of hydrogen bonds and in the process weakening the silk polymer. Elongation at break is about 20-25% and extension at break is 33% at 100% RH.

| Table: Properties of silk | |
|---|---|
| Properties | |
| Moisture regain | 11% |
| Shrinkage (wet) | 0.9% |
| Specific gravity | 1.3 |
| Tenacity | 5gr/denier |
| Elongation: dry | 17%-25% |
| wet | 30% |
| Rigidity modulus | 2.5 |

## Elastic – plastic Nature

Very crystalline polymer system does not permit the amount of polymer movement, which results in plastic nature of silk than elastic. Hence, if silk textile material is stretched, the silk polymers, which have beta-configuration, will slide past each other. This stretching results in rupturing of hydrogen bonds. After stretching, the polymers do not return to their original position, which leads to distortion and wrinkling or creasing of the silk textile material.

## Effect of Heat

Heating of silk fiber remains unaffected for a long period at 1400C. It is more sensitive to heat compared to wool, which is due to the lack of any covalent cross links in the polymer system of silk, compared with the di-sulphide bonds which are present in wool's polymer system. Peptide bonds, salt-linkages and hydrogen bonds of silk fiber decompose quickly at 175°C.

## Effect of UV Radiation

The strength and elongation of silk fiber decreases when fiber is exposed to UV radiation. But degree of crystallinity is not affected by the radiation.

## Chemical Properties

## Effect of Acids and Bases

Acids and alkalis cause hydrolysis of polypeptide chain present in the fibroin fiber. Acid hydrolysis causes more damage to fiber than alkaline hydrolysis. Acid hydrolysis attacks at nearly all the peptide linkages as compared to alkali hydrolysis, which occurs at the end of the peptide chain. Concentrated sulphuric acid and hydrochloric acid will dissolve the fiber and nitric acid result in change of color of silk. Dilute acid does not attack the fiber under mild condition. Hot caustic alkalis readily dissolve the fiber. Weak alkalis attack fibroin when the treatment time at boiling point is prolonged.

## Effect of Organic Solvents

Organic solvents are used as dry cleaning solvents and they do not dissolve the silk fiber.

## Effect of Oxidizing Agents

Oxidation reactions are considered to take place at the side chains of tyrosine, the amino terminal residues of the main chains and at the peptide bonds. Silk fiber is susceptible to oxidizing agents. Therefore care is required during bleaching. Hydrogen peroxide and some per acids are used in bleaching of raw silk.

## Effect of Reducing Agents

Fibroin resists the reducing agents such as hydrosulphite, sulphurous acid and its salt.

## Color-fastness

The luster of silk will cause its dyed and printed silk textile materials to appear much brighter in color.

## Applications of Fiber

Property of silk to absorb water easily makes it comfortable to wear and hence widely used for clothing. Its attractive lustre and drape makes it appropriate for applications such as home furnishing. It has potential to use as biomedical application as surgical sutures as it doesn't cause inflammatory reactions and biodegradable microtubes for repair of blood vessels. Apart from these it is used in cosmetic, pharmaceutical and dietary applications.

## Shellac

Shellac is the commercial resin marketed in the form of amber flakes, made from lac, the secretion of the family of lac-producing insects, though most commonly from the cultivated Kerria lacca, found in the forests of India and Thailand.

Shellac is obtained from the secretion of the female insect, harvested from the bark of the trees where she deposits it to provide a sticky hold on the trunk. There is a risk that the harvesting process can scoop the insect up along with the secretion, leading to its death. The natural coloration of lac residue is greatly influenced by the sap consumed by the lac insect and the season of the harvest.

## Properties

Shellac is soluble in alkaline solutions such as ammonia, sodium borate, sodium carbonate, and sodium hydroxide, and also in various organic solvents. When dissolved in alcohol, typically blends containing ethanol and methanol, shellac yields a coating of superior durability and hardness and is available in numerous grades. It is used in the traditional "French polish" method of finishing furniture, and fine viols and guitars. Shellac is also used as a finish for certain former Soviet Bloc small arms' wood stocks, such as the stock of the AK-47. Shellac refined for industrial purposes either retains its natural wax content or is refined wax-free by filtration. Orange shellac is bleached with sodium hypochlorite solution to form white shellac and also is produced in wax-containing and wax-free form. Because it is compatible with most other finishes, shellac is also used as a barrier or primer coat on wood to prevent the bleeding of resin or pigments into the final finish, or to prevent wood stain from blotching. Lightly tinted shellac preparations are also sold as paint primer. Shellac is best suited to this application because, although its durability against abrasives and many common solvents is not very good, it provides an excellent barrier against water vapor penetration. Shellac based primers are thus an effective sealant to control odors associated with fire damage.

## Natural Rubber

Natural rubber is an addition polymer that is obtained as a milky white fluid known as latex from a tropical rubber tree. Natural rubber is from the monomer isoprene

(2-methyl-1, 3-butadiene), which is a conjugated diene hydrocarbon as mentioned above. In natural rubber, most of the double fond formed in the polymer chain have the Z configuration, resulting in natural rubber's elastomer qualities.

## Production of Natural Rubber

The raw material from which natural rubber is made comes from the sap of rubber trees. The rubber plants are tapped for collecting the rubber latex. For this, an incision is made into the bark of the rubber tree and the latex sap is collected in cups. After collecting the latex sap, the raw natural rubber is refined to convert it into a usable rubber. Initially an acid was added to the latex, which used to make the sap set like a jelly. The latex jelly thus obtained was then flattened and rolled into rubber sheets and hung out to dry. In the year 1839, Charles Goodyear invented a more sophisticated way of making rubber stronger and more elastic. This was the process of rubber vulcanizing. The unprocessed natural rubber is sticky, deforms easily when warm, and is brittle when cold. In such a state, it cannot be used to make products having a good level of elasticity. Vulcanization prevents the polymer chains from moving independently. As a result, when stress is applied the vulcanized rubber deforms, but upon release of the stress, the product reverts to its original shape.

## Source of Natural Rubber

The natural rubber is produced from hundreds of different plant species. However, the most important source is from a tropical tree known as Hevea brasiliensis, which is native to the tropical Americas. This tree grows best in areas with an annual rainfall of just under 2000mm and at temperatures of 21-28 degrees. Due to these features and the preferred altitude of the tree around 600 meters, the prime growing area is around 10 degrees on either side of the equator.

## Properties of Natural Rubber

Natural rubber has certain unique properties such as follows:

- Natural rubber combines high strength (tensile and tear) with outstanding resistance to fatigue.

- It has excellent green strength and tack, which means that it has the ability to stick to itself and to other materials, which makes it easier to fabricate.

- It has moderate resistance to environmental damage by heat, light and ozone, which is one of its drawback.

- The natural rubber has excellent adhesion to brass-plated steel cord, which is ideal in rubber tyres.

- It has low hysteresis, which leads to low heat generation, and this in turn maintains new tyre service integrity and extends retread ability.

- The natural rubber has low rolling resistance with enhanced fuel economy.

- It has high resistance to cutting, chipping and tearing.

## Uses of Natural Rubber

- Natural rubber forms an excellent barrier to water.

- This is possibly the best barrier against pathogens such as the AIDS virus (HIV). That is the reason why latex is used in in condoms and surgical and medical examination gloves.

- Natural rubber is an excellent spring material.

- Natural rubber latex is also used in catheters, balloons, medical tubes, elastic thread, and also in some adhesives.

- Other than rayon, it is the sole raw material, which is used by the automotive industry.

- Rubber wood is another byproduct of natural rubber, which is growing in importance. It is a source of charcoal for local cooking.

## Cellulose

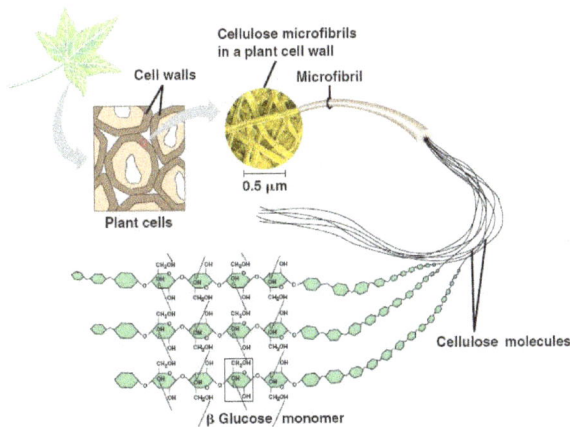

Cellulose is the most abundant polysaccharide found in nature. It is a linear polymer consisting of 6-member ether rings (D-glucose or dextrose) linked together covalently by ether groups, the so-called glycosidic bonds. Usually many thousand glucose repeat units make up a cellulose polymer. Cellulose and its derivatives can be considered condensation polymers because their hydrolysis yields glucose molecules:

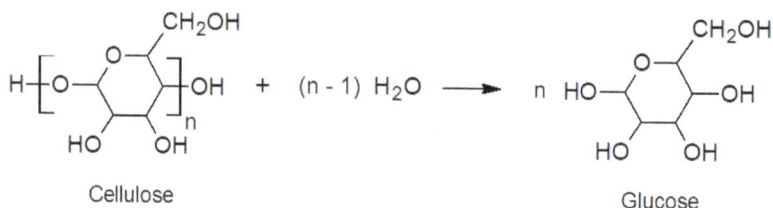

Cellulose                                    Glucose

The cyclic structure in the main polymer chain together with strong hydrogen bonding gives cellulose a rigid structure. Thus, cellulose and some of its derivatives have a high glass transition temperature and melting point. The strong intermolecular hydrogen bonds between the hydroxyl groups lead to highly ordered crystalline regions with low accessibility to reactants, which explains why cellulose is water insoluble and why strong alkalis like caustic soda are required to break down the structure to make the hydroxyl groups accessible to reactants.

Cellulose is the most abundant organic polymer on the planet. It is an important structural component of the primary cell wall of plants. The cellulose content of cotton fibers is about 90 percent. Not surprising, it is the main raw material for many semi-synthetic cellulose derivatives.

## Ester Cellulose

The most important cellulose esters are cellulose acetate (CAc), and the co-esters cellulose acetate-propionate, and cellulose acetate-butyrate. Among these, cellulose acetate is by far the most important cellulose ester. It was first used for photographic film and later as a coating for fabric on airplanes. Like cellophane it is made from cellulose but has very different properties. Unlike cellophane, it is thermoplastic, that is, it will soften and melt when heated.

The most common source of cellulose is cotton linters. The fibers are mixed with glacial acetic acid and acetic anhydride with sulfuric acid as a catalyst. This results in cellulose triacetate. In a subsequent step water is added to stop the reaction and to partially hydrolyze the triacetate:

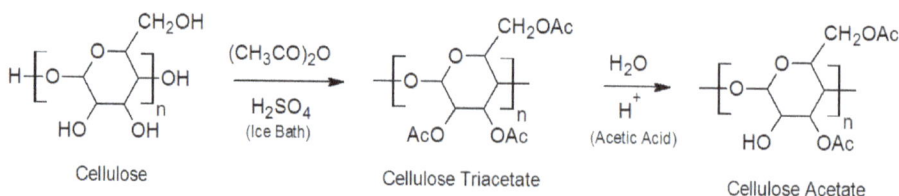

Cellulose                     Cellulose Triacetate                     Cellulose Acetate

Cellulose acetate is a crystal clear, tough, and flexible plastic and is the most stable cellulose derivative. It has excellent chemical resistance to organic and inorganic weak acids, hydrocarbons, vegetable oils, and the like. Often plasticizers are added to further increase its flexibility or mixed ester of cellulose like butyrate-acetate and propionate-acetate are produced which have improved flexibility, toughness, and moisture resistance.

## Ether Cellulose

Cellulose ethers are produced from wood pulp or cotton linters. The cellulose is treated with a solution of sodium hydroxide in a process similar to cellophane. In a subsequent step, the alkali cellulose is treated with an alkyl halide or an epoxide. The first method is frequently used to prepare ethyl cellulose whereas the second method is used to prepare and hydroxethyl and hydroxpropyl cellulose. Alternatively, the alkali cellulose can be treated with alkyl sulfate. For example, methyl sulfate treatment is a common process for the manufacture of methyl cellulose.

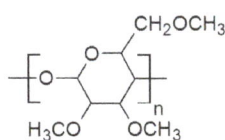

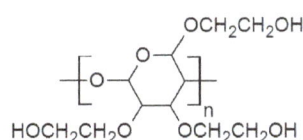

The most important modified cellulose polymers are methyl cellulose (MC), and ethyl cellulose (EC).

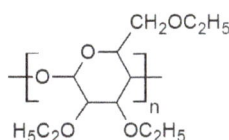

Ethyl Cellulose                Methyl Cellulose                Hydroxyethyl Cellulose

Other commercially important cellulose ethers include hydroxylpropyl cellulose (HPC), hydroxyethyl cellulose (HEC), and carboxymethyl cellulose CMC. These polymers can be produced by treating alkali cellulose with epoxides (HPC, HEC) or with chloroacetate (CMC).

Methyl cellulose (MC) is the most important commercial cellulose ether. It is also the simplest derivative where methoxy groups have replaced the hydroxyl groups. The most important properties of this nonionic polymer are its water solubility and its gelation when exposed to heat. Although soluble in water, films made from methyl cellulose usually retain their strength and do not become tacky when exposed to humidity. Polymer films made of methyl cellulose have excellent strength (60 - 70 MPa) and low elongation (5 - 15 %) at room temperature (75°F) but its strength decreases rapidly with increasing temperature. MC also has excellent UV, oil, and solvent resistance.

Ethyl cellulose (EC) is another important commercial cellulose ether derivative. While complete etherification is possible yielding triethyl cellulose usually only to 2 to 2.5 ethoxyl groups per glucose unit are etherified. This polymer has excellent strength at room temperature but its strength decreases rapidly with increasing temperature.

The lower the number of ether groups the greater will be the toughness and the lower the solubility, but poorer compatibility with plasticizers and other additives will result.

Partially hydrolyzed cellulose ethers and esters can also be converted to thermoset resins. The crosslinking can be achieved by reacting the residual hydroxyl groups with urea formaldehydes, melamines, or polyisocyantes.

## Cellulose Nitrate (Celluloid)

Nitrocellulose (NC), also called cellulose nitrate, is the oldest thermoplastic. It was invented by Alexander Parkes in 1855 and later commercialized under the trademarks *Parkesine*, *Xylonite* and *Celluloid*. To achieve the desired properties, other additives such as camphor, dyes, stabilizers and fillers are added.

Cellulose nitrate itself is synthesized by mixing cellulose fibers with an aqueous solution of nitric and sulfuric acid. The fibers are immersed in this solution for 20 to 60 minutes at 30 to 40°C. The product is then repeatedly washed with water and sodium carbonate solution to neutralized and removes the acids.

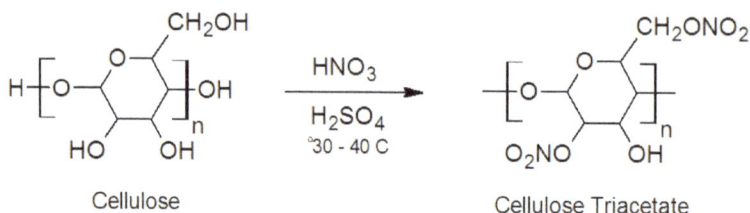

Cellulose                                      Cellulose Triacetate

The average degree of nitration will be affected by the water content, composition of the bath, immersion time, and reaction condition. NC's with about 2 nitrate groups per glucose repeat unit are often chosen in plastics and laquers. A higher nitrate content is used in explosives.

Cellulose nitrate has excellent mechanical properties. However, plastics made from NC like celluloid have poor weathering and heat resistance and are not resistant to dilute acids and bases, but are insoluble and stable in water and nonpolar solvents.

Nitrocellulose is highly combustible which makes it too hazardous for most applications. Today, NC is mainly used as a binder in products like inks, coatings, and adhesives. The dilution with other ingredients greatly reduces its flammability.

## Commercial Cellulose Esters and Ethers

Major manufacturers of cellulose esters and ethers are AkzoNobel (CMC, EHEC, MEHEC), Ashland (MC, HPMC, CMC), Dow Chemical (MC, HPMC, HEC, EC, NC), Eastman (CAc), Daicel (CAc), Mitsubishi Rayon (Regenerated Cellulose), Shin Etsu (MC, HPMC), Solvay (CAc), and Tembec (CAc, MC).

## Applications

Cellulose is mainly used to produce paper and paperboard. Only relative small quantities are converted to semi-synthetic cellulose derivatives, such as cellophane, rayon, and cellulose acetate and cellulose ethers.

The most important cellulose ester is *cellulose acetate*. It is widely used for industrial applications and can be classified into two types: cellulose diacetate and cellulose tri-acetate. Important uses include textiles (fibers and threads for quality fabrics); plastic films such as optical film for LCD technology and antifog goggles; and consumer products such as cellulose based filters, window cartons, and labels.

The main application of *methyl cellulose* is water-soluble films used for packaging products that dissolve in water like medical capsules, bubble bath, tooth pastes, detergent powders, rat poison and bread dough. Other important uses of methyl cellulose include ceramic tile adhesive and grout formulations, wallpaper adhesives, shampoos, cosmetics and a number of other products where thermal gelation, viscosity adjustment, and water solubility is required. In these formulations it functions as a thickener, binder, film former, and/or as a water retention agent. For example, methylcellulose together with ethyl, carboxymethyl, and hydroxethyl cellulose is used as a thickener in many food products.

*Ethyl cellulose* (EC) is mainly used as a plastic film similarly to cellulose acetate. It is not as widely used as methyl cellulose because there are cheaper plastics that have similar properties. EC can be used for similar applications as methyl cellulose. Like MC, it functions as a binder, water barrier, rheology modifier, and suspension stabilizer in various applications.

Another important cellulose ether is *hydroxyethyl cellulose*. It too finds applications as a thickener, binder, stabilizer, and film former and as a protective colloid. Like MC, it is used in the manufacture of wallboards, as a component of emulsions for surface treatment to improve adhesion of surface coatings, as moisture retaining agent and retarder for cement formulation and as a thickening agent in wallpaper pastes.

*Nitrocellulose* (NC) is used on a much smaller scale, predominantly as a binder in printing inks, glues, wood coatings, paints and laquers (for musical instruments). Celluloid plastics are used on an even smaller scale. Major applications include table tennis balls, filter membranes and celluloid film (until the 1930th). A big drawback of cellulose nitrate is its flammability. For this reason, it has been replaced by cellulose acetate and vinyl polymers in most plastic applications.

## Synthetic Polymers

Synthetic polymers are those which are human-made polymers. Polymers are those, which consists of repeated structural units known as monomers. Polyethylene is considered to be as one of the simplest polymer, it has ethane or ethylene as the monomer unit whereas the linear polymer is known as the high density polyethylene-HDPE. Many of the polymeric materials have chain-like structures, which resemble polyethylene.

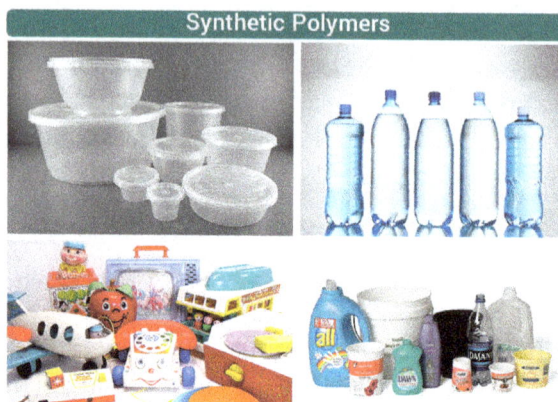

Synthetic polymers are sometimes referred as "plastics", of which the well-known ones are nylon and polyethylene. The polymers, which are formed by linking monomer units, without the any change of material, are known to as addition polymers or also called as chain-growth polymers. All these are said to be synthetic polymers.

Some of the synthetic polymers, which we use in our everyday life, include nylons used in fabrics and textiles, Teflon used in non-stick pans, polyvinyl chloride used in pipes. The PET bottles we use are commonly made up of synthetic polymer called as polyethylene terephthalate. The covers and plastic kits comprise of synthetic polymers such as polythene, and the tyres of vehicles are manufactured from the Buna rubbers. But on the other side, there also arises environmental issues by the use of these synthetic polymers such as the bio plastics and those made from petroleum as they are said to be non-biodegradable.

## Types of Synthetic Polymers

### Low Density Polyethylene

Low Density Polyethylene (LDPE) polymers are among the most common types of synthetic organic polymers, which are often found in households. LDPE is a thermoplastic made from the monomer ethylene. One of the first polymers to be created, it was produced in 1933 by Imperial Chemical Industries using a high-pressure process via free radical polymerization. It is manufactured the way method today. LDPE is commonly recycled, with the number 4 as its recycling symbol. Despite competition from more modern polymers, LDPE continues to be an important plastic grade.

### High Density Polyethylene

High Density Polyethylene (HDPE) or polyethylene high-density (PEHD) is a polyethylene thermoplastic made from petroleum. It takes 1.75 kilograms of petroleum (in terms of energy and raw materials) to make one kilogram of HDPE. HDPE is commonly recycled, with the number 2 as its recycling symbol.

## Polypropylene

Polypropylene (PP), also known as polypropene, is a thermoplastic polymer used in a wide variety of applications, including packaging and labeling, textiles, stationery, plastic parts and reusable containers of various types, laboratory equipment, loudspeakers, automotive components, and polymer banknotes. An additional polymer made from the monomer propylene, it is rugged and unusually resistant to many chemical solvents, bases, and acids.

## Polyvinyl Chloride

Polyvinyl Chloride (PVC) is the third-most widely produced plastic, after polyethylene and polypropylene. PVC is used in construction because it is cheaper and stronger than more traditional alternatives such as copper or ductile iron. It can be made softer and more flexible by adding plasticizers, the most popular of which are phthalates. In this form, PVC is used in clothing and upholstery, electrical cable insulation, inflatable products, and many applications in which it replaces rubber.

## Polystyrene

Polystyrene (PS) is an aromatic polymer made from the monomer styrene, a liquid petrochemical. One of the most popular plastics, PS is a colorless solid that is used, for example, in disposable cutlery, plastic models, CD and DVD cases, and smoke detector housings. Products made from foamed polystyrene include packing materials, insulation, and foam drink cups. Its very slow biodegradation is a focus of controversy, and it can often be found littered outdoors, particularly along shores and waterways.

## Nylon

Nylon, a family of synthetic polymers known generically as polyamides. Nylon is one of the most commonly used polymers. The amide backbone present in nylon causes it to be more hydrophilic than the polymers discussed above. Notice that your nylon clothing will absorb water, for instance; this is because nylon can engage in hydrogen bonding with water, unlike the purely hydrocarbon polymers that make up most plastics.

## Teflon

Teflon (Polytetrafluoroethylene or PTFE) is a synthetic fluoropolymer of tetrafluoroethylene, and has numerous applications. PTFE is a solid, high-molecular-weight compound consisting entirely of carbon and fluorine. PTFE is hydrophobic: neither water nor water-containing substances can interact with PTFE. PTFE is used as a nonstick coating for pans and other cookware because it has very low friction with other compounds. It is very non-reactive, partly because of the strength of carbon–fluorine bonds, so it is often used in containers and pipework for reactive and corrosive chemicals. Where used as a lubricant, PTFE reduces friction, wear, and energy consumption of machinery.

Teflon frying pan Teflon (PTFE) is often used to coat non-stick frying pans as it is hydrophobic and possesses fairly high heat resistance.

## Thermoplastic Polyurethane

Thermoplastic polyurethane (TPU) is any of a class of polyurethane plastic. It has many useful properties, including elasticity, transparency, and resistance to oil, grease, and abrasion. Most of these properties are resultant of the fact that TPU is hydrophilic and can react with water. Technically, TPU is a thermoplastic elastomer consisting of linear segmented block copolymers made of hard and soft segments.

## Synthetic Polymers- Uses

Some of the uses are given below-

1.   The polymer called Polyethylene is used in plastic bags and film wraps.

2.   Polyethylene is utilized in the bottles, electrical insulation, toys, etc.

3. Polyvinyl Chloride (PVC) is used in siding, pipes, flooring purposes.

4. The synthetic polymer Polystyrene is used in cabinets and in packaging.

5. Polyvinyl acetate is used in adhesives and latex paints.

## Thermoplastic Polymer

In thermoplastics polymers, monomer units are linked by intermolecular interactions like Van Der Waals forces. The monomer units are arranged in linear or branched manner.

The arrangement of monomer units in thermoplastic material can be compared to a set of strings, which are mixed on a table. Here each string is represents a polymer. As the degree of mixing increases, it requires great effort to separate them.

## Thermoplastic Polymer

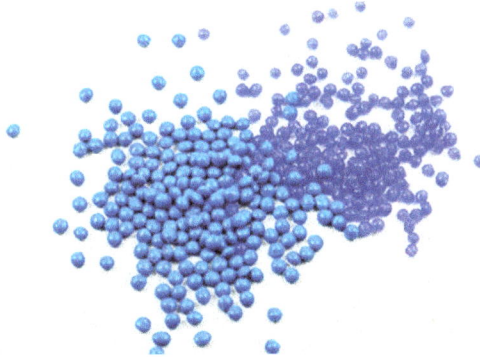

Thermoplastic polymers can be defined as the polymers, which get soften while heating and can be remolded in different shapes. On the basis of degree of the intermolecular interactions between the polymer chains, the polymer can be classified as amorphous or crystalline forms. Both forms can exist with thermoplastic polymers.

The amorphous structure acquires a bundled structure therefore such structures are responsible for elastic properties of polymers. On the contrary, crystal structures of polymeric chains contain an ordered and compacted structure like lamellar structures and micellar form. Such type of structures is associated with mechanical properties and the temperature resistance of polymers. The thermoplastic polymer with high concentration of amorphous structures is a poor resistance to loads with an excellent elasticity. Similarly thermoplastic polymer with high concentration of crystalline structure is very strong and shows little elasticity.

## Thermoplastic Polymers Properties

These polymers melt by heating and may melt even before passing to a gaseous state. They can deform while heating and soluble in certain solvents. In some of the solvents, they swell. Thermoplastic polymers show good resistance to creep. These polymers soften when heated and change to fluid. This process is completely reversible because no chemical bonding takes place during this process. Due to this property, such polymers can be remolded and recycled. Most of the thermoplastic polymers show high strength, shrink-resistance and easy bendability.

They are highly recyclable, have high-impact resistance, remolding and reshaping capabilities, chemical resistant, hard crystalline or rubbery surface. High cost and easy melting are common limitations of thermoplastic polymers. Amorphous thermoplastic polymers are transparent or translucent, with low tendency to creep, good dimensional stability, low tendency to warp, high brittleness, low chemical resistance and very sensitive to stress cracking. Semi-crystalline thermoplastic polymers do not exhibit a clear rubbery region. Polyamides are most common examples of such polymers. Some transition in amorphous thermoplastic polymer can form semi-crystalline polymers.

## Thermoplastic Polymers Examples

High-pressure polyethylene, low-pressure polyethylene, polystyrene, polyamide and PVC or polyvinyl chlorides are some common examples of thermoplastic polymers. Some common examples of thermoplastic adhesives are Acrylates, Cyanoacrylates, Epoxy and Acrylates cured by ultraviolet radiation.

High-pressure polyethylene is mainly used to cover rigid material like electrical machines, tubes, etc. On the contrary, low-pressure polyethylene is an elastic material that is mainly used as insulator in electrical cables. Polystyrene is one of the common thermoplastic polymers that are applied for electrical insulation and as handles of tools. Polyamide is used for the manufacturing of ropes and belts. PVC or polyvinyl chloride is mainly used for the manufacturing of insulation materials, pipes and containers.

## Elastomer

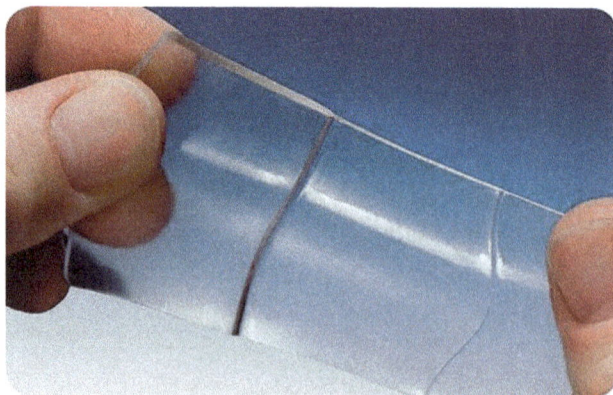

An elastomer is a polymer with the property of elasticity. In other words, it is a polymer that deforms under stress and returns to its original shape when the stress is removed.

The term is a contraction of the words "elastic polymer." There are many types of elastomers, most of which are rubbers. The term elastomer is therefore often used interchangeably with the term rubber. Other elastomers, which melt when heated, are classified as thermoplastic.

**Thermoplastic**     **Elastomer**     **Thermoset**

Rubbers (both natural and synthetic) are widely used for the manufacture of tires, tubes, hoses, belts, matting, gloves, toy balloons, rubber bands, pencil erasers, and adhesives. Thermoplastic elastomers are used in manufacturing processes, such as by injection molding. Thermoplastic polyurethanes are used for various applications, including the production of foam seating, seals, gaskets, and carpet underlay.

## Properties of Elastomer

Elastomers are amorphous polymers with considerable segmental motion. Their general molecular form has been likened to a "spaghetti and meatball" structure, where the meatballs signify cross-links between the flexible polymer chains, which are like spaghetti strands.

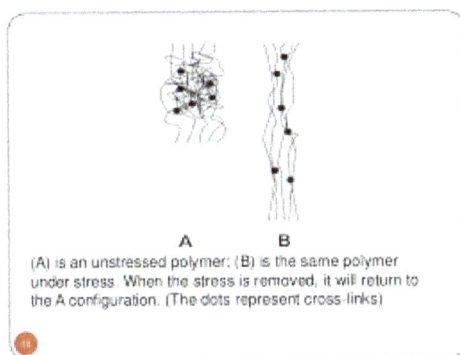

A     B
(A) is an unstressed polymer; (B) is the same polymer under stress. When the stress is removed, it will return to the A configuration. (The dots represent cross-links)

Each polymer chain is made up of many monomer subunits, and each monomer is usually made of carbon, hydrogen, and oxygen atoms, and occasionally silicon atoms.

Most elastomers are thermosets—that is, they require curing (by heat, chemical reaction, or irradiation). In the curing process, the long polymer chains become cross-linked by covalent bonds, the material becomes stronger, and it cannot be remelted and remolded.

Some elastomers are thermoplastic, melting to a liquid state when heated and turning brittle when cooled sufficiently. In thermoplastic elastomers, the polymer chains are cross-linked by weaker bonds, such as hydrogen bonds or dipole-dipole interactions.

The elasticity is derived from the ability of the long chains to reconfigure themselves to distribute an applied stress. Covalent cross-linkages, in particular, ensure that the elastomer will return to its original configuration when the stress is removed.

The temperature of the polymer also affects its elasticity. Elastomers that have been cooled to a glassy or crystalline phase will have less mobile chains, and consequently less elasticity, than those manipulated at temperatures higher than the glass transition temperature of the polymer.

At ambient temperatures, rubbers are thus relatively soft (Young's modulus of about 3 MPa) and deformable.

## Examples of Elastomers

Unsaturated rubbers that can be cured by sulfur vulcanization:

- Natural Rubber (NR)

- Synthetic Polyisoprene (IR)

- Butyl rubber (copolymer of isobutylene and isoprene, IIR)

  - Halogenated butyl rubbers (Chloro Butyl Rubber: CIIR; Bromo Butyl Rubber: BIIR)

- Polybutadiene (BR)

- Styrene-butadiene Rubber (copolymer of polystyrene and polybutadiene, SBR)

- Nitrile Rubber (copolymer of polybutadiene and acrylonitrile, NBR), also called Buna N rubbers

  - Hydrogenated Nitrile Rubbers (HNBR) Therban and Zetpol

- Chloroprene Rubber (CR), polychloroprene, Neoprene and Baypren etc.

  Saturated Rubbers that cannot be cured by sulfur vulcanization:

- EPM (ethylene propylene rubber, a copolymer of ethylene and propylene) and EPDM rubber (ethylene propylene diene rubber a terpolymer of ethylene, propylene and a diene-component)

- Epichlorohydrin rubber (ECO)

- Polyacrylic rubber (ACM, ABR)

- Silicone rubber (SI, Q, VMQ)

- Fluorosilicone Rubber (FVMQ)

- Fluoroelastomers (FKM, and FEPM) Viton, Tecnoon, Fluorel, Aflas and Dai-El

## Various other types of Elastomers

- Thermoplastic elastomers (TPE), for example Elastron, etc.

- Thermoplastic Vulcanizates (TPV), for example Santoprene TPV

- Thermoplastic Polyurethane (TPU)

- Thermoplastic Olens (TPO)

- The proteins resilin and elastin

- Polysulde Rubber

## Uses

Most elastomers are rubbers, including both natural and synthetic varieties. They are used mainly for the manufacture of tires and tubes. They are also used to produce goods such as hoses, belts, matting, gloves, toy balloons, rubber bands, pencil erasers, and adhesives. As a fiber, rubber (called "elastic") is valuable for the textile industry.

Ethylene propylene rubber (EPR) is useful as insulation for high voltage cables. Thermoplastic elastomers are relatively easy to use in manufacturing processes, such as by injection molding.

Polyurethanes are widely used in high-resiliency, flexible foam seating, seals, gaskets, carpet underlay, Spandex fibers, and electrical potting compounds.

## Mathematical Background

Using the laws of thermodynamics, stress definitions, and polymer characteristics, ideal stress behavior may be calculated using the following equation:

$$\sigma = nkT\left[\lambda_1^2 + \lambda_1^{-1}\right]$$

where, $n$ is the number of chain segments per unit volume, $k$ is Boltzmann's Constant, $T$ is temperature, and is distortion in the $\lambda_1$ direction is distortion in the 1 direction.

## Thermosetting Polymer

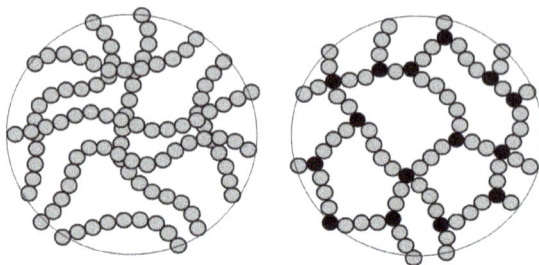

Left: individual linear polymer chains
Right: Polymer chains which have being cross-linked to give a rigid 3D thermoset polymer

A thermosetting polymer (also called a thermosetting plastic or thermosetting resin) is a polymer, which becomes irreversibly hardened upon being cured. Curing is caused by the action of heat or suitable radiation and may be promoted by high pressure or the use of a catalyst. It results in extensive cross-linking between polymer chains to give an infusible and insoluble polymer network. A cured thermosetting polymer is called a thermoset.

Thermosetting resins are usually malleable or liquid prior to curing, and are often designed to be molded into their final shape, or used as adhesives. Once hardened a thermoset resin cannot be re-melted in order to be reshaped, this can be contrasted with thermoplastic polymers, which can generally be re-melted and reshaped.

## Examples

- Polyester resin fiberglass systems: sheet molding compounds and bulk molding compounds; filament winding; wet lay-up lamination; repair compounds and protective coatings.

- Polyurethanes: insulating foams, mattresses, coatings, adhesives, car parts, print rollers, shoe soles, flooring, synthetic fibers, etc. Polyurethane polymers are formed by combining two bi- or higher functional monomers/oligomers.

- Polyurea/polyurethane hybrids used for abrasion resistant waterproofing coatings.

- Vulcanized rubber.

- Bakelite, a phenol-formaldehyde resin used in electrical insulators and plasticware.

- Duroplast, light but strong material, similar to Bakelite used for making car parts.

- Urea-formaldehyde foam used in plywood, particleboard and medium-density fiberboard.

- Melamine resin used on worktop surfaces.

- Diallyl-phthalate (DAP) used in high temperature and mil-spec electrical connectors and other components. Usually glass filled.

- Epoxy resin used as the matrix component in many fiber reinforced plastics such as glass-reinforced plastic and graphite-reinforced plastic; casting; electronics encapsulation; construction; protective coatings; adhesives; sealing and joining.

- Epoxy novolac resins used for printed circuit boards, electrical encapsulation, adhesives and coatings for metal.

- Benzoxazines, used alone or hybridised with epoxy and phenolic resins, for structural prepregs, liquid molding and film adhesives for composite construction, bonding and repair.

- Polyimides and Bismaleimides used in printed circuit boards and in body parts of modern aircraft, aerospace composite structures, as a coating material and for glass reinforced pipes.

- Cyanate esters or polycyanurates for electronics applications with need for dielectric properties and high glass temperature requirements in aerospace structural composite components.

- Mold or mold runners (the black plastic part in integrated circuits or semiconductors).

- Furan resins used in the manufacture of sustainable biocomposite construction, cements, adhesives, coatings and casting/foundry resins.

- Silicone resins used for thermoset polymer matrix composites and as ceramic matrix composite precursors.

- Thiolyte, an electrical insulating thermoset phenolic laminate material.

- Vinyl ester resins used for wet lay-up laminating, molding and fast setting industrial protection and repair materials.

**Thermosetting Polymers**

❑A thermosetting polymer is one which becomes hard on heating and it cannot be softened by heating again.

## Curing

Curing a thermosetting resin transforms it into a plastic, or elastomer (rubber) by crosslinking or chain extension through the formation of covalent bonds between individual chains of the polymer. Crosslink density varies depending on the monomer or prepolymer mix, and the mechanism of crosslinking:

Acrylic resins, polyesters and vinyl esters with unsaturated sites at the ends or on the backbone are generally linked by copolymerisation with unsaturated monomerdiluents, with cure initiated by free radicals generated from ionizing radiation or by the photolytic or thermal decomposition of a radical initiator – the intensity of crosslinking is influenced by the degree of backbone unsaturation in the prepolymer.

Epoxy functional resins can be homo-polymerized with anionic or cationic catalysts and heat, or copolymerised through nucleophilic addition reactions with multifunctional crosslinking agents which are also known as curing agents or hardeners. As reaction proceeds, larger and larger molecules are formed and highly branched crosslinked structures develop, the rate of cure being influenced by the physical form and functionality of epoxy resins and curing agents – elevated temperature postcuring induces secondary crosslinking of backbone hydroxyl functionality which condense to form ether bonds.

Polyurethanes form when isocyanate resins and prepolymers are combined with low- or high-molecular weight polyols, with strict stochiometric ratios being essential to control nucleophilic addition polymerisation – the degree of crosslinking and resulting physical type (elastomer or plastic) is adjusted from the molecular weight and functionality of isocyanate resins, prepolymers, and the exact combinations of diols, triols and polyols selected, with the rate of reaction being strongly influenced by catalysts and inhibitors; polyureas form virtually instantaneously when isocyanate resins are combined with long-chain amine functional polyether or polyester resins and short-chain diamine extenders – the amine-isocyanate nucleophilic addition reaction does not require catalysts. Polyureas also form when isocyanate resins come into contact with moisture.

Phenolic, amino and furan resins all cure by polycondensation involving the release of water and heat, with cure initiation and polymerisation exotherm control influenced by curing temperature, catalyst selection/loading and processing method/pressure – the degree of pre-polymerisation and level of residual hydroxymethyl content in the resins determine the crosslink density.

## Properties

Thermosetting plastics are generally stronger than thermoplastic materials due to the three-dimensional network of bonds (crosslinking), and are also better suited to high-temperature applications up to the decomposition temperature since they keep their shape, as strong covalent bonds between polymer chains cannot be easily broken. The higher the crosslink density and aromatic content of a thermoset polymer, the higher the resistance to heat degradation and chemical attack. Mechanical strength and hardness also improve with crosslink density, although at the expense of brittleness.

Thermoset plastic polymers characterized by rigid, three-dimensional structures and high molecular weight, stay out of shape when deformed and undergo permanent or plastic deformation under load, and normally decompose before melting. Thermoset elastomers, which are soft and springy or rubbery and can be deformed and revert to their original shape on loading release, also decompose before melting. Conventional thermoset plastics or elastomers therefore cannot be melted and re-shaped after they are cured which implies that thermosets cannot be recycled for the same purpose, except as filler material. There are developments however involving thermoset epoxy resins, which on controlled and contained heating form cross-linked networks that can be repeatedly reshaped like silica glass by reversible covalent bond exchange reactions on reheating above the glass transition temperature. There are also thermoset polyurethanes shown to have transient properties and which can thus be reprocessed or recycled.

Thermosetting polymer mixtures based on thermosetting resin monomers and pre-polymers can be formulated and applied and processed in a variety of ways to create distinctive cured properties that cannot be achieved with thermoplastic polymers or inorganic materials. Application/process uses and methods for thermosets include protective coating, seamless flooring, civil engineering construction grouts for jointing and injection, mortars, foundry sands, adhesives, sealants, castings, potting, electrical insulation, encapsulation, 3D printing, solid foams, wet lay-up laminating, pultrusion, gelcoats, filament winding, pre-pregs, and molding. Specific methods of molding thermosets are:

- Reactive injection moulding (used for objects such as milk bottle crates).

- Extrusion molding (used for making pipes, threads of fabric and insulation for electrical cables).

- Compression molding (used to shape SMC and BMC thermosetting plastics).

- Spin casting (used for producing fishing lures and jigs, gaming miniatures, figurines, emblems as well as production and replacement parts).

## Fiber-reinforced Materials

When compounded with fibers thermosetting resins form fiber-reinforced polymer composites, which are used in the fabrication of factory finished structural composite OEM or replacement parts, and as site-applied, cured and finished composite repair and protection materials. When used as the binder for aggregates and other solid fillers, they form particulate-reinforced polymer composites, which are used for factory-applied protective coating or component manufacture, and for site-applied and cured construction, or maintenance, repair and overhaul (MRO) purposes.

## Synthetic Rubber

Synthetic rubber is a man-made rubber, which is produced, in manufacturing plants by synthesizing it from petroleum and other minerals. Synthetic rubber is basically a polymer or an artificial polymer. It has the property of undergoing elastic stretch ability or deformation under stress but can also return to its previous size without permanent deformation.

## Synthetic Rubber Production

### Polymerization Methods

Synthetic elastomers are produced on an industrial scale in either solution or emulsion polymerization methods. Polymers made in solution generally have more linear molecules (that is, less branching of side chains from the main polymer chain), and they also have a narrower distribution of molecular weight (that is, greater length) and flow more easily. In addition, the placement of the monomer units in the polymer molecule can be controlled more precisely when polymerization is conducted in solution. The monomer or monomers are dissolved in a hydrocarbon solvent, usually hexane or cyclohexane, and polymerized, using an organometallic catalyst such as butyllithium.

In emulsion polymerization, the monomer (or monomers) is emulsified in water with a suitable soap (e.g., sodium stearate) employed as a surfactant, and a water-soluble free-radical catalyst (e.g., potassium per sulfate, peroxides, a redox system) is added to induce polymerization. After polymerization has reached the desired level, the reaction is stopped by adding a radical inhibitor. About 10 percent of synthetic elastomer produced through emulsion techniques is sold as latex. The rest is coagulated with acidified brine, washed, dried, and pressed into 35-kg (77-pound) bales.

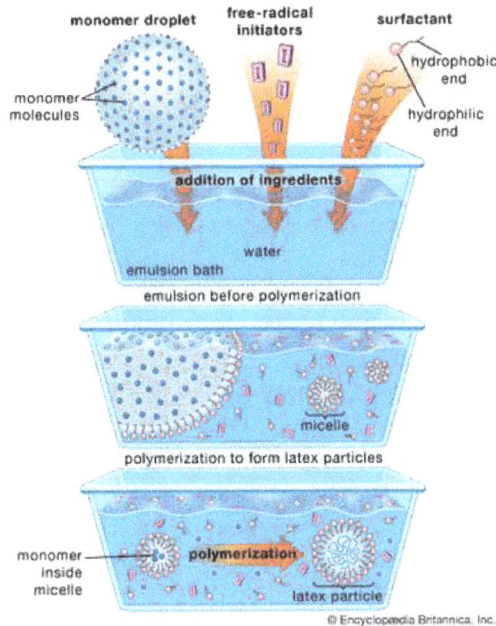

Schematic diagram of the emulsion-polymerization method.

Monomer molecules and free-radical initiators are added to a water-based emulsion bath along with soap like materials known as surfactants, or surface-acting agents. The surfactant molecules, composed of a hydrophilic (water-attracting) and hydrophobic (water-repelling) end, form a stabilizing emulsion before polymerization by coating the monomer droplets. Other surfactant molecules clump together into smaller aggregates called micelles, which also absorb monomer molecules. Polymerization occurs when initiators migrate into the micelles, inducing the monomer molecules to form large molecules that make up the latex particle.

When emulsion polymerization of SBR is carried out "hot" (i.e., at 50° C, or 120° F), the polymer molecules are more branched. When polymerization is carried out "cold" (i.e., at 5° C, or 40° F), they are more linear and generally higher in molecular weight—features that improve the rolling resistance and wear resistance of tires. In some cases polymerization is continued in order to give products of such high molecular weight that they would normally be intractable. In these cases about 30 percent of a heavy oil is added before coagulation to yield "oil-extended" elastomers with superior wear resistance.

## Chemical Types of Synthetic Rubber

Synthetic rubber is artificially made from petrochemical feed stocks. Crude oil is the principal raw material for different types of rubber in synthetic category. As opposed to natural rubber where there is only one chemical type, there are approximately 20 different chemical types of synthetic rubber, and within all of the types of rubber, there are different grades. The different types of rubber, especially the synthetic rubber types have their own individual properties and advantages. The industry chooses the rubber types, which most clearly meet the demands of an intended use.

## Types and uses of Synthetic Rubber

Various types of rubbers have been synthesized since the invention of synthetic rubber. Given below are some of the common types of synthetic rubbers that are used in different industries:

- Polychloroprene (CR): It is also commonly known as 'neoprene' and shows a greater resistance to heat along with better chemical stability. Owing to these properties, it is used in laptop sleeves, gaskets, fan belts of automobiles, and hoses.

- Styrene-Butadiene (SBR): This rubber shows better resistance to abrasion as well as wear and tear, and is hence used in tires, mainly of buses and aircraft. It is also used in conveyor belts and the soles of shoes.

- Ethylene Propylene Diene Monomer (EPDM): Along with heat and weather, this rubber shows good resistance to various chemicals. It is used in the heat collectors present in solar panels, mechanical vibrators, electrical insulation, and radiators.

- Acrylonitrile Butadiene (NBR): It shows better resistance to chemicals. This makes it useful in the production of lab gloves and oil seals. It is also used in synthetic leather, V belts, and O rings.

- Polysiloxane (SI): It provides electrical insulation, and has low chemical and thermal conductivity. Also known as silicone rubber, it is used in coatings, as a sealant, and to make molds like the ones used in dentistry.

- Chloro Isobutylene Isoprene (CIIR): It has good physical properties, and shows resistance to heat and weathering. It is used as a additive in oils and fuels. It is also used in the manufacturing of various sports goods, and chewing gum as well.

- Chlorosulphonated Polyethylene (CSM): Along with resistance to chemicals and temperature, this rubber also is resistant to UV light. It is used in coating as well roofing materials, and foldable kayaks.

- Perfluoroelastomer (FFKM): This rubber has good resistance to chemicals and temperature. It is used in the fabrication of silicon wafers. It is also used in chemical processing and high-pressure seals.

- Thermoplastic Polyether-ester (YBPO): This rubber has high flexibility, and shows good resistance to chemicals. It is used as buffers, and in the production of belts and moldings.

## Others Uses

Apart from the uses mentioned above, some other uses of synthetic rubber include:

- Production of weather balloons.
- In mattresses and pillows.
- In fuel for launching rockets during WWII.
- As it is waterproof, it is used in the manufacturing inflatable boats and diving suits.

Initially, rubber was used to rub out the marks of a pencil, and from here it derived its name. It is known as 'Indian Rubber'.

## Synthetic Fiber

Synthetic fiber is a man-made textile fiber produced entirely from chemical substances, unlike those man-made fibers derived from such natural substances as cellulose or protein.

Synthetic fibers, on the other hand, undergo changes in their chemical structure and composition, during the manufacturing process. Polymers such as regenerated cellulose, polycaprolactam, and polyethylene terephthalate, which have become familiar household materials under the trade names, Rayon, Nylon, and Dacron, respectively, are also made into numerous non fiber products, ranging from cellophane envelope windows to clear plastic soft-drink bottles. As fibers, these materials are prized for their strength, toughness, resistance to heat and mildew, and ability to hold a pressed form.

## Types of Synthetic Fibers

1. Polyester is made from esters of dihydric alcohol and terpthalic acid.

2. Acrylic fabrics are polycrylonitriles.

3. Rayon is recycled wood pulp that is treated with chemicals like caustic soda, ammonia, acetone and sulphuric acid to survive regular washing and wearing.

4. Acetate and Triacetate are made from wood fibers called cellulose and undergo extensive chemical processing to produce the finished product.

5. Nylon is made from petroleum and is often given a permanent chemical finish that can be harmful.

## Uses of synthetic fibers

Synthetic fibers play an important role in today's world and are used either on their own or mixed with other synthetic or natural fibers to create fabrics or products for everyday use. Some uses are:

1.  Ropes

2.  Parachutes

3.  Fish Nets

4.  Carpets

5.  Tents

6.  Fillers in pillows

7.  Fabrics for everyday wear like lycra and spandex

8.  Blankets

9.  Warm and protective clothing for extreme climates

10. Synthetic hair wigs

## Advantages of Synthetic Fibers

Synthetic fibers are used because of their durable nature. Some of the advantages are:

*   They have good elasticity.

*   They do not wrinkle easily.

*   They are comparatively less expensive, more durable, require less maintenance and are easily available.

*   They are stronger and can handle heavy loads.

## Disadvantages of Synthetic Fibers

*   Most are not heat resistant making them dangerous to wear near fire.

*   They do not allow air circulation, making them sticky, sweaty and uncomfortable to wear, during hot and humid climates.

*   They are non – biodegradable.

## Nylon

Nylon is an umbrella term for synthetic materials that can be processed into different shapes and textures to be put to various uses. Nylon is a plastic with super-long, heavy molecules built up of short, continually repeating sections of atoms. The polymers can be mixed with various substances to achieve different variations in properties, which explain why this material has such a varied use.

## Production of Nylon

Nylon is made by reacting together two large molecules, diamine acid and dicarboxylic acid, which fuse together to make an even larger molecule and give off water. The large polymer formed in this case is the most common type of nylon called nylon-6, 6. A giant sheet or ribbon of nylon is produced that is shredded into chips, which becomes the raw material for plastic products. Nylon textiles are made from fibers of nylon, by melting nylon chips and running them through a wheel with several tiny holes in it. Fibers of different length and thickness are made by using holes of different size and drawing them out at different speeds.

## Using of Nylon used

One of the common uses of nylon is to make women's stockings and hosiery. The nylon blends are used for making swimwear, track pants, windbreakers etc. Other uses include parachutes, umbrellas, luggage, netting for veils etc. Because of its resistance to heat and cold, strong and lightweight nature, it's also used to make ropes, such as the ones used in boats.

## Advantages of Nylon

- Lightweight
- Strong
- Resistant to Abrasion
- Easy to wash and dry
- Resists shrinkage and wrinkle

- Water-proof
- Cost-effective

## Disadvantages of Nylon

- Can spark due to static charge
- Low absorbency
- Melts if catches fire

## Polyester

Polyester is a term often defined as "long-chain polymers chemically composed of at least 85% by weight of an ester and a dihydric alcohol and a terephthalic acid". In other words, it means the linking of several esters within the fibers. Reaction of alcohol with carboxylic acid results in the formation of esters.

Polyester also refers to the various polymers in which the backbones are formed by the "esterification condensation of polyfunctional alcohols and acids".

Polyester can also be classified as saturated and unsaturated polyesters.

Saturated polyesters refer to that family of polyesters in which the polyester backbones are saturated. They are thus not as reactive as unsaturated polyesters. They consist of low molecular weight liquids used as plasticizers and as reactants in forming urethane polymers, and linear, high molecular weight thermoplastics such as polyethylene terephthalate. Usual reactants for the saturated polyesters are a glycol and an acid or anhydride.

Unsaturated polyesters refer to that family of polyesters in which the backbone consists of alkyl thermosetting resins characterized by vinyl unsaturation. They are mostly used in reinforced plastics. These are the most widely used and economical family of resins.

## Characteristics of Polyester

- Polyester fabrics and fibers are extremely strong.

- Polyester is very durable: resistant to most chemicals, stretching and shrinking, wrinkle resistant, mildew and abrasion resistant.

- Polyester is hydrophobic in nature and quick drying. It can be used for insulation by manufacturing hollow fibers.

- Polyester retains its shape and hence is good for making outdoor clothing for harsh climates.

- It is easily washed and dried.

## Industry

## Basics

Polyester is a synthetic polymer made of purified terephthalic acid (PTA) or its dimethyl ester dimethyl terephthalate (DMT) and monoethylene glycol (MEG). With 18% market share of all plastic materials produced, it ranges third after polyethylene (33.5%) and polypropylene (19.5%).

The main raw materials are described as follows:

Purified terephthalic acid (PTA) CAS-No.: 100-21-0

Synonym: 1,4 benzenedicarboxylic acid,

Sum formula: $C_6H_4(COOH)_2$, mol. weight: 166.13

Dimethylterephthalate (DMT) CAS-No.: 120-61-6

Synonym: 1,4 benzenedicarboxylic acid dimethyl ester,

Sum formula: $C_6H_4(COOCH_3)_2$, mol. weight: 194.19

Mono-ethylene glycol (MEG) CAS No.: 107-21-1

Synonym: 1,2 ethanediol,

Sum formula: $C_2H_6O_2$ , mol. weight: 62.07

To make a polymer of high molecular weight a catalyst is needed. The most common catalyst is antimony trioxide (or antimony tri-acetate):

Antimony trioxide (ATO) CAS-No.: 1309-64-4

mol. weight: 291.51,

Sum formula: $Sb_2O_3$

In 2008, about 10,000 tonnes $Sb_2O_3$ were used to produce around 49 million tonnes polyethylene terephthalate.

Polyester is described as follows:

Polyethylene terephthalate CAS-No.: 25038-59-9

Synonyms/abbreviations: polyester, PET, PES,

Sum formula: $H-[C_{10}H_8O_4]-n=60-120$ OH, mol. unit weight: 192.17

There are several reasons for the importance of polyester:

- The relatively easy accessible raw materials PTA or DMT and MEG.

- The very well understood and described simple chemical process of polyester synthesis.

- The low toxicity level of all raw materials and side products during polyester production and processing.

- The possibility to produce PET in a closed loop at low emissions to the environment.

- The outstanding mechanical and chemical properties of polyester.

- The recyclability.

- The wide variety of intermediate and final products made of polyester.

In the following table, the estimated world polyester production is shown. Main applications are textile polyester, bottle polyester resin, film polyester mainly for packaging and specialty polyesters for engineering plastics. According to this table, the world's total polyester production might exceed 50 million tons per annum before the year 2010.

| World polyester production by year | | |
|---|---|---|
| Product type | 2002 (million tonnes/year) | 2008 (million tonnes/year) |
| Textile-PET | 20 | 39 |
| Resin, bottle/A-PET | 9 | 16 |
| Film-PET | 1.2 | 1.5 |
| Special polyester | 1 | 2.5 |
| Total | 31.2 | 59 |

## Raw Material Producer

The raw materials PTA, DMT, and MEG are mainly produced by large chemical companies which are sometimes integrated down to the crude oil refinery where p-Xylene is the base material to produce PTA and liquefied petroleum gas (LPG) is the base material to produce MEG.

## Polyester Processing

After the first stage of polymer production in the melt phase, the product stream divides into two different application areas, which are mainly textile applications and packaging applications. In the following table, the main applications of textile and packaging of polyester are listed.

| Textile and packaging polyester application list (melt or pellet) | |
|---|---|
| Textile | Packaging |
| Staple fiber (PSF) | Bottles for CSD, water, beer, juice, detergents, etc. |
| Filaments POY, DTY, FDY | A-PET film |
| Technical yarn and tire cord | Thermoforming |
| Non-woven and spunbond | biaxial-oriented film (BO-PET) |
| Mono-filament | Strapping |

Abbreviations:

PSF

Polyester-staple fiber:

POY

Partially oriented yarn:

DTY

Drawn textured yarn:

FDY

Fully drawn yarn:

CSD

Carbonated soft drink:

A-PET

Amorphous polyester film:

BO-PET

Biaxial-oriented polyester film:

A comparable small market segment (much less than 1 million tones/year) of polyester is used to produce engineering plastics and master batch.

In order to produce the polyester melt with a high efficiency, high-output processing steps like staple fiber (50–300 tones/day per spinning line) or POY /FDY (up to 600 tones/day split into about 10 spinning machines) are meanwhile more and more vertically integrated direct processes. This means the polymer melt is directly converted into the textile fibers or filaments without the common step of pelletizing. We are talking about full vertical integration when polyester is produced at one site starting from crude oil or distillation products in the chain oil → benzene → PX → PTA → PET melt → fiber/filament or bottle-grade resin. Such integrated processes are meanwhile established in more or less interrupted processes at one production site. Eastman Chemicals were the first to introduce the idea of closing the chain from PX to PET resin with their so-called INTEGREX process. The capacity of such vertically integrated production sites is >1000 tones/day and can easily reach 2500 tones/day.

Besides the above-mentioned large processing units to produce staple fiber or yarns, there are ten thousands of small and very small processing plants, so that one can estimate that polyester is processed and recycled in more than 10 000 plants around the globe. This is without counting all the companies involved in the supply industry, beginning with engineering and processing machines and ending with special additives, stabilizers and colors. This is a gigantic industry complex and it is still growing by 4–8% per year, depending on the world region.

## Synthesis

Synthesis of polyesters is generally achieved by a polycondensation reaction. The general equation for the reaction of a diol with a diacid is:

$$(n+1) \; R(OH)_2 + n \; R'(COOH)_2 \rightarrow HO[ROOCR'COO]_n ROH + 2n \; H_2O$$

## Azeotrope Esterification

In this classical method, an alcohol and a carboxylic acid react to form a carboxylic ester. To assemble a polymer, the water formed by the reaction must be continually removed by azeotrope distillation.

## Alcoholic Transesterification

Transesterification: An alcohol-terminated oligomer and an ester-terminated oligomer condense to form an ester linkage, with loss of an alcohol. R and R' are the two oligomer chains, R" is a sacrificial unit such as a methyl group (methanol is the byproduct of the esterification reaction).

## Acylation (HCl Method)

The acid begins as an acid chloride, and thus the polycondensation proceeds with emission of hydrochloric acid (HCl) instead of water. This method can be carried out in solution or as an enamel.

## Silyl Method

In this variant of the HCl method, the carboxylic acid chloride is converted with the trimethyl silyl ether of the alcohol component and production of trimethyl silyl chloride is obtained.

## Ring-opening Polymerization

Aliphatic polyesters can be assembled from lactones under very mild conditions, catalyzed anionically, cationically or metallorganically. A number of catalytic methods for the copolymerization of epoxides with cyclic anhydrides have also recently been shown to provide a wide array of functionalized polyesters, both saturated and unsaturated.

## Biodegradation

The futuro house was made of fibreglass-reinforced polyester plastic; polyester-polyurethane, and poly(methylmethacrylate) one of them was found to be degrading by Cyanobacteria and Archaea.

## Cross-linking

Unsaturated polyesters are thermosetting resins. They are generally copolymers prepared by polymerizing one or more diol with saturated and unsaturated dicarboxylic acids (maleic acid, fumaric acid) or their anhydrides. The double bond of unsaturated polyesters reacts with a vinyl monomer, usually styrene, resulting in a 3-D cross-linked structure. This structure acts as a thermoset. The exothermic cross-linking reaction is initiated through a catalyst, usually an organic peroxide such as methyl ethyl ketone peroxide or benzoyl peroxide.

## Acrylic Fiber

Acrylic fiber is a synthetic fiber that closely resembles wool in its character. According to the definition of the ISO (International Standards Organization) and BISFA

(International Synthetic Fiber Standardization Office), fibers, which contain a minimum of 85% acrylonitrile in their chemical structure are called "Acrylic Fibers".

Acrylic fiber is composed of acrylonitrile and a comonomer. The comonomer is added to improve dyeability and the textile processability of the acrylic fiber. Acrylic fiber is produced with two different systems: wet spinning and dry spinning. Acrylic fiber can be supplied as producer-dyed either by pigmentation of the dope or with jel dyeing systems. It can be used 100% alone, or in blends with other natural and synthetic fibers.

Basic Principles of Acrylic Fiber Production — Acrylic fibers are produced from acrylonitrile, a petrochemical. The acrylonitrile is usually combined with small amounts of other chemicals to improve the ability of the resulting fiber to absorb dyes. Some acrylic fibers are dry spun and others are wet spun. Acrylic fibers are used in staple or tow form.

Acrylic fibers are modified to give special properties best suited for particular end-uses. They are unique among synthetic fibers because they have an uneven surface, even when extruded from a round-hole spinneret.

## Acrylic Fiber Characteristics

- Outstanding wick ability & quick drying to move moisture from body surface.
- Flexible aesthetics for wool-like, cotton-like, or blended appearance.
- Easily washed, retains shape.
- Resistant to moths, oil, and chemicals.
- Dye able to bright shades with excellent fastness.
- Superior resistance to sunlight degradation.

## Some Major Acrylic Fiber Uses

- Apparel: Sweaters, socks, fleece wear, circular knit apparel, sportswear and children's wear.

- Home Furnishings: Blankets, area rugs, upholstery, pile; luggage, awnings, outdoor furniture.

- Other Uses: Craft yarns, sail cover cloth, wipe cloths.

- Industrial Uses: Asbestos replacement; concrete and stucco reinforcement.

## General Acrylic Fiber Care Tips

- Wash delicate items by hand in warm water. Static electricity may be reduced by using a fabric softener in every third or fourth washing. Gently squeeze out water, smooth or shake out garment and let dry on a non-rust hanger. *(Sweaters, however, should be dried flat).*

- When machine washing, use warm water and add a fabric softener during the final rinse cycle.

- Machine dry at a low temperature setting. Remove garments from dryer as soon as tumbling cycle is completed.

- If ironing is required, use a moderately warm iron.

# Bioplastic

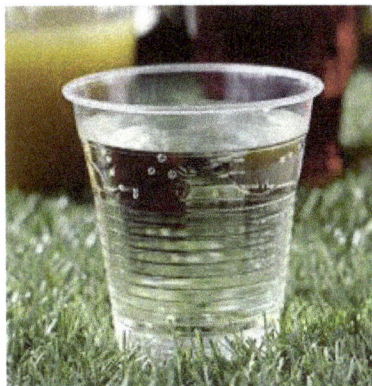

A bioplastic is a substance made from organic biomass sources, unlike conventional plastics, which are made from petroleum.

Bioplastics are made through a number of different processes. Some use a microorganism to process base materials, such as vegetable oils, cellulose, starches, acids and alcohols.

While almost all bioplastics produce less carbon dioxide in production than conventional plastics, they are not necessarily completely green. The methods by which their base materials are grown and the processing involved both impact their product footprint. Many bioplastics also release carbon dioxide or monoxide when biodegrading. Nevertheless, their overall environmental impact is typically lower than that of conventional plastics, and as oil costs rise, their cost becomes more and more competitive.

Some biodegradable bioplastics can break down in 180 days, given the right conditions. Others are not biodegradable at all. This capacity is desirable, for example, for outdoors applications where longevity and a reduced carbon footprint in production may be the goals.

Bioplastics, like petroleum-based ones, differ in make up to address different needs. The bioplastics used to make disposable cutlery, food containers, grocery bags, electronics casings and conductive bioplastics for electronics are all very different from one another.

## Properties of Bioplastic

- Bio-based and biodegradable bioplastics: These are made using renewable resources, such as plant biomass, and will biodegrade under certain environmental conditions. These materials are suitable for disposable items, such as packaging, drink bottles, single-use food containers and cutlery. They are more sustainable because they save fossil fuel resources and, if disposed of appropriately, support further plant growth.

- Bio-based and durable (non-biodegradable) bioplastics: These are made using renewable resources but are designed to have a longer life span (for example, carpet fibers and interior car panels). Using renewable resources makes these materials more sustainable. Also, using them to replace metal components in vehicles has the advantage of reducing vehicle weight, which increases fuel efficiency.

- Petrochemical-based and biodegradable bioplastics: There are some petrochemical-based plastics that can be biodegraded by the microbes in the soil, compost or oceans.

## Types

## Starch-based Plastics

Thermoplastic starch currently represents the most widely used bioplastic, constituting about 50 percent of the bioplastics market. Simple starch bioplastic can be made at home. Pure starch is able to absorb humidity, and is thus a suitable material for the production of drug capsules by the pharmaceutical sector. Flexibiliser and plasticiser such as sorbitoland glycerine can also be added so the starch can also be processed thermo-plastically. The characteristics of the resulting bioplastic (also called "thermo-plastical starch") can be tailored to specific needs by adjusting the amounts of these additives.

Starch-based bioplastics are often blended with biodegradable polyesters to produce starch/polylactic acid,starch/polycaprolactone or starch/Ecoflex (polybutylene adipate-co-terephthalate produced by BASF). blends. These blends are used for industrial applications and are also compostable. Other producers, such as Roquette, have developed other starch/polyolefin blends. These blends are not biodegradable, but have a lower carbon footprint than petroleum-based plastics used for the same applications.

## Cellulose-based Plastics

A packaging blister made from cellulose acetate, a bioplastic

Cellulose bioplastics are mainly the cellulose esters, (including cellulose acetate and nitrocellulose) and their derivatives, including celluloid.

## Protein-based Plastics

Bioplastics can be made from proteins from different sources. For example, wheat gluten and casein show promising properties as a raw material for different biodegradable polymers.

## Some aliphatic Polyesters

The aliphatic biopolyesters are mainly polyhydroxyalkanoates (PHAs) like the poly-3-hydroxybutyrate (PHB), polyhydroxyvalerate (PHV) and polyhydroxyhexanoate (PHH).

## Polylactic Acid (PLA)

Mulch film made of polylactic acid(PLA)-blend bio-flex

Polylactic acid (PLA) is a transparent plastic produced from corn or dextrose. Superficially, it is similar to conventional petrochemical-based mass plastics like PS. It has the distinct advantage of degrading to nontoxic products. Unfortunately it exhibits inferior impact strength, thermal robustness, and barrier properties (blocking air transport across the membrane). PLA and PLA blends generally come in the form of granulates with various properties, and are used in the plastic processing industry for the production of films, fibers, plastic containers, cups and bottles. PLA is also the most common type of plastic filament used for home fused deposition modeling.

## Poly-3-hydroxybutyrate

The biopolymer poly-3-hydroxybutyrate (PHB) is a polyester produced by certain bacteria processing glucose, corn starch or wastewater. Its characteristics are similar to those of the petroplastic polypropylene. PHB production is increasing. The South American sugar industry, for example, has decided to expand PHB production to an industrial scale. PHB is distinguished primarily by its physical characteristics. It can be processed into a transparent film with a melting point higher than 130 degrees Celsius, and is biodegradable without residue.

## Polyhydroxyalkanoates

Polyhydroxyalkanoates are linear polyesters produced in nature by bacterial fermentation of sugar or lipids. They are produced by the bacteria to store carbon and energy. In industrial production, the polyester is extracted and purified from the bacteria by optimizing the conditions for the fermentation of sugar. More than 150 different monomers can be combined within this family to give materials with extremely different properties. PHA is more ductile and less elastic than other plastics, and it is also biodegradable. These plastics are being widely used in the medical industry.

## Polyamide 11

PA 11 is a biopolymer derived from natural oil. It is also known under the tradename Rilsan B, commercialized by Arkema. PA 11 belongs to the technical polymers family and is not biodegradable. Its properties are similar to those of PA 12, although emissions of greenhouse gases and consumption of nonrenewable resources are reduced during its production. Its thermal resistance is also superior to that of PA 12. It is used in high-performance applications like automotive fuel lines, pneumatic airbrake tubing, electrical cable antitermite sheathing, flexible oil and gas pipes, control fluid umbilicals, sports shoes, electronic device components, and catheters.

A similar plastic is Polyamide 410 (PA 410), derived 70% from castor oil, under the trade name EcoPaXX, commercialized by DSM. PA 410 is a high-performance polyamide that combines the benefits of a high melting point (approx. 250° C), low moisture absorption and excellent resistance to various chemical substances.

## Bio-derived Polyethylene

The basic building block (monomer) of polyethylene is ethylene. Ethylene is chemically similar to, and can be derived from ethanol, which can be produced by fermentation of agricultural feed stocks such as sugar cane or corn. Bio-derived polyethylene is chemically and physically identical to traditional polyethylene – it does not biodegrade but can be recycled. The Brazilian chemicals group Braskem claims that using its method of producing polyethylene from sugar cane ethanol captures (removes from the environment) 2.15 tones of $CO_2$ per tonne of Green Polyethylene produced.

## Genetically Modified Bioplastics

Genetic modification (GM) is also a challenge for the bioplastics industry. None of the currently available bioplastics – which can be considered first generation products – require the use of GM crops, although GM corn is the standard feedstock.

Looking further ahead, some of the second generation bioplastics manufacturing technologies under development employ the "plant factory" model, using genetically modified crops or genetically modified bacteria to optimize efficiency.

## Polyhydroxyurethanes

Recently, there have been a large emphasis on producing bio based and isocyanate-free polyurethanes. One such example utilizes a spontaneous reaction between polyamines and cyclic carbonates to produce polyhydroxurethanes. Unlike traditional cross-linked polyurethanes, cross-linked polyhydroxyurethanes have been shown to be capable of recycling and reprocessing through dynamic transcarbamoylation reactions.

## Lipid Derived Polymers

A number bioplastic classes have been synthesized from plant and animal derived fats and oils. Polyurethanes, polyesters, epoxy resins and a number of other types of polymers have been developed with comparable properties to crude oil based materials. The recent development of olefin metathesis has opened a wide variety of feed stocks to economical conversion into biomonomers and polymers. With the growing production of traditional vegetable oils as well as low cost microalgae derived oils, there is huge potential for growth in this area.

## Environmental Impact

Confectionery packaging made of PLA-blend bio-flex     Bottles made from cellulose acetatebiograde

The environmental impact of bioplastics is often debated, as there are many different metrics for "greenness" (e.g., water use, energy use, deforestation, biodegradation, etc.) and tradeoffs often exist. The debate is also complicated by the fact that many different types of bioplastics exist, each with different environmental strengths and weaknesses, so not all bioplastics can be treated as equal.

The production and use of bioplastics is sometimes regarded as a more sustainable activity when compared with plastic production from petroleum (petroplastic), because it requires less fossil fuel for its production and also introduces fewer, net-new greenhouse emissions if it biodegrades. The use of bioplastics can also result in less hazardous waste than oil-derived plastics, which remain solid for hundreds of years.

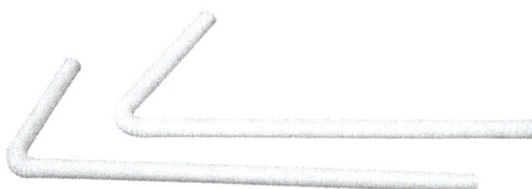

Drinking straws made of PLA-blend bio-flex

Jar made of PLA-blend bio-flex, a bioplastic

Petroleum is often still used as a source of materials and energy in the production of bioplastic. Petroleum is required to power farm machinery, to irrigate crops, to produce fertilizers and pesticides, to transport crops and crop products to processing plants, to process raw materials, and ultimately to produce the bioplastic. However, it is possible to produce bioplastic using renewable energy sources and avoid the use of petroleum.

Italian bioplastic manufacturer Novamont states in its own environmental audit that producing one kilogram of its starch-based product uses 500 g of petroleum and consumes almost 80% of the energy required to produce a traditional poly-ethylene polymer. Environmental data from Nature Works, the only commercial manufacturer of PLA (polylactic acid) bioplastic, says that making its plastic material delivers a fossil fuel saving of between 25 and 68 per cent compared with polyethylene, in part due to its purchasing of renewable energy certificates for its manufacturing plant.

A detailed study examining the process of manufacturing a number of common packaging items from traditional plastics and polylactic acid carried out by Franklin Associates and published by the Athena Institute shows that using bioplastic has a lower environmental impact for some products, and a higher environmental impact for others. This study, however, does not factor in the end-of-life environmental impact of these products, including possible methane emissions from landfills due to biodegradable plastics.

While production of most bioplastics results in reduced carbon dioxide emissions compared to traditional alternatives, there is concern that the creation of a global bioeconomy required to produce bioplastic in large quantities could contribute to an accelerated rate of deforestation and soil erosion, and could adversely affect water supplies. Careful management of a global bioeconomy would be required.

Other studies showed that bioplastics result in a 42% reduction in carbon footprint.

On October 21, 2010, a group of scientists reported that corn-based plastic ranked higher in environmental defects than the main products it replaces, such as HDPE, LDPE and PP. In the study, the production of corn-based plastics created more acidification, carcinogens, ecotoxicity, eutrophication, ozone depletion, respiratory effects and smog than the synthetic-based plastics they replaced. However the study also concluded that biopolymers trumped the other plastics for biodegradability, low toxicity, and use of renewable resources.

The American Carbon Registry has also released reports of nitrous oxide caused from corn growing which is 310 times more potent as a greenhouse gas than $CO_2$. Pesticides are also used in growing corn-based plastic.

## Biodegradation

Packaging air pillow made of PLA-blend bio-flex

The terminology used in the bioplastics sector is sometimes misleading. Most in the industry use the term bioplastic to mean a plastic produced from a biological source. All (bio- and petroleum-based) plastics are technically biodegradable, meaning they can be degraded by microbes under suitable conditions. However, many degrade so slowly that they are considered non-biodegradable. Some petrochemical-based plastics are considered biodegradable, and may be used as an additive to improve the performance of commercial bioplastics. Non-biodegradable bioplastics are referred to as durable. The biodegradability of bioplastics depends on temperature, polymer stability, and available oxygen content. The European standard EN 13432, published by the International Organization for Standardization, defines how quickly and to what extent a plastic must be degraded under the tightly controlled and aggressive conditions (at or above 140 °F (60 °C)) of an industrial composting unit for it to be considered biodegradable. This standard is recognized in many countries, including all of Europe, Japan and the US. However, it applies only to industrial composting units and does not set out a standard for home composting. Most bioplastics (e.g. PH) only biodegrade quickly in industrial composting units. These materials do not biodegrade quickly in ordinary compost piles or in the soil/water. Starch-based bioplastics are an exception, and will biodegrade in normal composting conditions.

The term "biodegradable plastic" has also been used by producers of specially modified petrochemical-based plastics that appear to biodegrade. Biodegradable plastic bag manufacturers that have misrepresented their product's biodegradability may now face legal action in the US state of California for the misleading use of the terms biodegradable or compostable. Traditional plastics such as polyethylene are degraded by ultra-violet (UV) light and oxygen. To prevent this, process manufacturers add stabilizing chemicals. However, with the addition of a degradation initiator to the plastic, it is possible to achieve a controlled UV/oxidation disintegration process. This type of plastic may be referred to as *degradable plastic* or *oxy-degradable plastic* or *photodegradable plastic* because the process is not initiated by microbial action. While some degradable plastics manufacturers argue that degraded plastic residue will be attacked by microbes, these degradable materials do not meet the requirements of the EN13432 commercial composting standard. The bioplastics industry has widely criticized oxo-biodegradable plastics, which the industry association says do not meet its requirements. Oxo-biodegradable plastics – known as "oxos" – are conventional petroleum-based products with some additives that initiate degradation. The ASTM standard for oxo-biodegradables is called the Standard Guide for Exposing and Testing Plastics that Degrade in the Environment by a Combination of Oxidation and Biodegradation (ASTM 6954). Both EN 13432 and ASTM 6400 are specifically designed for PLA and Starch based products and should not be used as a guide for oxos.

## Disposal Options for Bioplastics

With the expected increase in bioplastic products, it's important we understand the different categories of bioplastic and how best to dispose of each type – otherwise, they can contaminate waste systems.

- Recycling: Many plastics can be successfully recycled into new plastic materials at the end of their life. However, if plastics are not sorted into different categories, mixing some degradable plastics with other recycled plastics can reduce the performance and life span of the recycled product.

- Composting: Not all biodegradable products are suitable for composting. Products labeled 'compostable' must meet recognized international standards, which specify the composting conditions such as time, temperature and the environmental effects of the final compost. Also, industrial and home composting conditions are different.

- Landfill: Biodegradable and compostable plastics take a long time to degrade in the anaerobic environment of a landfill. Anaerobic digestion of organic materials also produces methane gas that, if not captured, adds to greenhouse gases in the atmosphere.

# Biopolymers

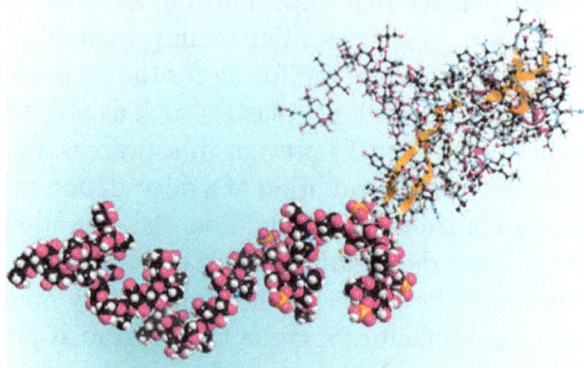

**Biopolymers**

Biopolymers are polymers that occur in nature. Carbohydrates and proteins, for example, are biopolymers. Many biopolymers are already being produced commercially on large scales, although they usually are not used for the production of plastics. Even if only a small percentage of the biopolymers already being produced were used in the production of plastics, it would significantly decrease our dependence on manufactured, non-renewable resources.

- Cellulose is the most plentiful carbohydrate in the world; 40 percent of *all organic matter* is cellulose!

- Starch is found in corn (maize), potatoes, wheat, tapioca (cassava), and some other plants. Annual world production of starch is well over 70 billion pounds, with much of it being used for non-food purposes, like making paper, cardboard, textile sizing, and adhesives.

- Collagen is the most abundant protein found in mammals. Gelatin is denatured collagen, and is used in sausage casings, capsules for drugs and vitamin preparations, and other miscellaneous industrial applications including photography.

- Casein, commercially produced mainly from cow's skimmed milk, is used in adhesives, binders, protective coatings, and other products.

- Soy protein and *zein* (from corn) are abundant plant proteins. They are used for making adhesives and coatings for paper and cardboard.

- Polyesters are produced by bacteria, and can be made commercially on large scales through fermentation processes. They are now being used in biomedical applications.

A number of other natural materials can be *made into polymers* that are biodegradable. For example:

- *Lactic acid* is now commercially produced on large scales through the fermentation of sugar feed stocks obtained from sugar beets or sugar cane, or from the conversion of starch from corn, potato peels, or other starch source. It can be polymerized to produce poly (lactic acid), which is already finding commercial applications in drug encapsulation and biodegradable medical devices.

- *Triglycerides* can also be polymerized. Triglycerides make up a large part of the storage lipids in animal and plant cells.

## Biopolymer Classification

There are four main types of Biopolymers. These are:

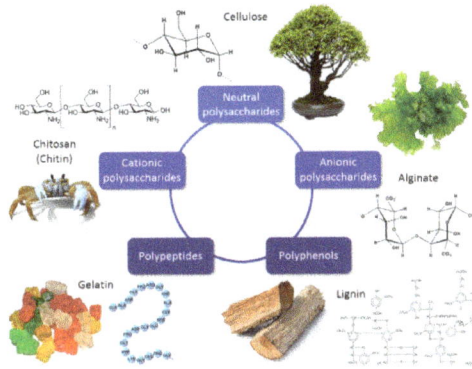

## 1. Sugar based Biopolymers

Starch or Sucrose is used as input for manufacturing Polyhydroxibutyrate. Sugar based polymers can be produced by blowing, injection, vacuum forming and extrusion. Lactic acid polymers (Polyactides) are created from milk sugar (lactose) that is extracted from potatoes, maize, wheat and sugar beet. Polyactides are resistant to water and can be manufactured by methods like vacuum forming, blowing and injection molding.

## 2. Starch based Biopolymers

Starch acts as a natural polymer and can be obtained from wheat, tapioca, maize and potatoes. The material is stored in tissues of plants as one way carbohydrates. It is composed of glucose and can be obtained by melting starch. This polymer is not present in animal tissues. It can be found in vegetables like tapioca, corn, wheat and potatoes.

## 3. Biopolymers based on Synthetic materials

Synthetic compounds that are obtained from petroleum can also be used for making biodegradable polymers such as aliphatic aromatic copolyesters. Though these polymers are manufactured from synthetic components, they are completely compostable and bio-degradable.

## 4. Cellulose based Biopolymers

These are used for packing cigarettes, CDS and confectionary. This polymer is composed of glucose and is the primary constituent of plant cellular walls. It is obtained from natural resources like cotton, wood, wheat and corn.

The production of biopolymer may be done either from animal products or agricultural plants.

## Biopolymer Types

There are primarily two types of Biopolymer, one that is obtained from living organisms and another that is produced from renewable resources but require polymerization. Those created by living beings include proteins and carbohydrates.

## Conventions and Nomenclature

## Polypeptides

The convention for a polypeptide is to list its constituent amino acid residues as they occur from the amino terminus to the carboxylic acid terminus. The amino acid residues are always joined by peptide bonds. Protein, though used colloquially to refer to any polypeptide, refers to larger or fully functional forms and can consist of several polypeptide chains as well as single chains. Proteins can also be modified to include non-peptide components, such as saccharide chains and lipids.

## Nucleic Acids

The convention for a nucleic acid sequence is to list the nucleotides as they occur from the 5' end to the 3' end of the polymer chain, where 5' and 3' refer to the numbering of carbons around the ribose ring which participate in forming the phosphate diester linkages of the chain. Such a sequence is called the primary structure of the biopolymer.

## Sugars

Sugar-based biopolymers are often difficult with regards to convention. Sugar polymers can be linear or branched and are typically joined with glycosidic bonds. The exact placement of the linkage can vary, and the orientation of the linking functional groups is also important, resulting in α- and β-glycosidic bonds with numbering definitive of the linking carbons' location in the ring. In addition, many saccharide units can undergo various chemical modifications, such as amination, and can even form parts of other molecules, such as glycoproteins.

## Structural Characterization

There are a number of biophysical techniques for determining sequence information.

Protein sequence can be determined by Edman degradation, in which the N-terminal residues are hydrolyzed from the chain one at a time, derivatized, and then identified. Mass spectrometer techniques can also be used. Nucleic acid sequence can be determined using gel electrophoresis and capillary electrophoresis. Lastly, mechanical properties of these biopolymers can often be measured using optical tweezers or atomic-force microscopy. Dual polarization interferometry can be used to measure the conformational changes or self-assembly of these materials when stimulated by pH, temperature, ionic strength or other binding partners.

## As Materials

Some biopolymers- such as PLA, naturally occurring zein, and poly-3-hydroxybutyrate can be used as plastics, replacing the need for polystyrene or polyethylene based plastics.

Some plastics are now referred to as being 'degradable', 'oxy-degradable' or 'UV-degradable'. This means that they break down when exposed to light or air, but these plastics are still primarily (as much as 98 per cent) oil-based and are not currently certified as 'biodegradable' under the European Union directive on Packaging and Packaging Waste (94/62/EC). Biopolymers will break down, and some are suitable for domestic composting.

Biopolymers (also called renewable polymers) are produced from biomass for use in the packaging industry. Biomass comes from crops such as sugar beet, potatoes or wheat: when used to produce biopolymers, these are classified as non food crops. These can be converted in the following pathways:

Sugar beet > Glyconic acid > Polyglyconic acid

Starch > (fermentation) > Lactic acid > Polylactic acid (PLA)

Biomass > (fermentation) > Bioethanol > Ethene > Polyethylene

Many types of packaging can be made from biopolymers: food trays, blown starch pellets for shipping fragile goods, thin films for wrapping.

## Environmental Impacts

Biopolymers can be sustainable, carbon neutral and are always renewable, because they are made from plant materials, which can be grown indefinitely. These plant materials come from agricultural non food crops. Therefore, the use of biopolymers would create a sustainable industry. In contrast, the feedstocks for polymers derived from petrochemicals will eventually deplete. In addition, biopolymers have the potential to cut carbon emissions and reduce $CO_2$ quantities in the atmosphere: this is because the $CO_2$ released when they degrade can be reabsorbed by crops grown to replace them: this makes them close to carbon neutral.

Biopolymers are biodegradable, and some are also compostable. Some biopolymers are biodegradable: they are broken down into $CO_2$ and water by microorganisms. Some of these biodegradable biopolymers are compostable: they can be put into an industrial composting process and will break down by 90% within six months. Biopolymers that do this can be marked with a 'compostable' symbol, under European Standard EN 13432 (2000). Packaging marked with this symbol can be put into industrial composting processes and will break down within six months or less. An example of a compostable polymer is PLA film under 20μm thick: films, which are thicker than that do not qualify as compostable, even though they are "biodegradable". In Europe there is a home composting standard and associated logo that enables consumers to identify and dispose of packaging in their compost heap.

## Biopolymer Uses

These polymers play an essential role in nature. They are extremely useful in performing functions like storage of energy, preservation and transmittance of genetic information and cellular construction.

- Sugar based polymers, such as Polyactides, naturally degenerate in the human body without producing any harmful side effects. This is the reason why they are used for medical purposes. Polyactides are commonly used as surgical implants.

- Starch based biopolymers can be used for creating conventional plastic by extruding and injection molding.

- Biopolymers based on synthetic are used to manufacture substrate mats.

- Cellulose based Biopolymers, such as cellophane, are used as a packaging material.

- These chemical compounds can be used to make thin wrapping films, food trays and pellets for sending fragile goods by shipping.

## Biopolymer Environmental Benefits

Some of the environmental benefits of this polymer are:

- These polymers are carbon neutral and can always be renewed. These are sustainable as they are composed of living materials.

- These polymers can reduce carbon dioxide levels in the atmosphere and also decrease carbon emissions. This happens because bio-degradation of these chemical compounds can release carbon dioxide that can be reabsorbed by crops grown as a substitute in their place.

- It is also compostable which means there is less chance of environmental pollution from this compound. This is one of the primary advantages of this chemical compound. However, the materials composed from this compound are not compostable.

- These chemical compounds reduce dependency on non-renewable fossil fuels. These are easily biodegradable and can decrease air pollution. It greatly reduces the harmful effect of plastic use on the environment. Long-term use of biopolymer use will limit the use of fossil fuel.

# References

- Polyurethane Handbook, ed. G Oertel, Hanser, Munich, Germany, 2nd edition, 1994, ISBN 1569901570, ISBN 978-1569901571

- Chemical-engineering/natural-polymers: sciencedirect.com, Retrieved 18 July 2018

- Williams, Alan. "Washing clothes releases thousands of microplastic particles into environment, study shows". Plymouth University. Retrieved 9 October 2016

- Physical-chemical-properties-wool-fibre: textileapex.blogspot.com, Retrieved 22 June 2018

- Napper, I. E.; Thompson, R. C. (2016). "Release of Synthetic Microplastic Plastic Fibres From Domestic Washing Machines: Effects of Fabric Type and Washing Conditions". Marine Pollution Bulletin. 112 (1–2): 39–45. doi:10.1016/j.marpolbul.2016.09.025. PMID 27686821

- Natural-rubber: industrialrubbergoods.com, Retrieved 22 May 2018

- Can, E.; Küsefoğlu, S.; Wool, R. P. (2001-07-05). "Rigid, thermosetting liquid molding resins from renewable resources. I. Synthesis and polymerization of soy oil monoglyceride maleates". Journal of Applied Polymer Science. 81 (1): 69–77. doi:10.1002/app.1414. ISSN 1097-4628

- Thermoplastic-polymers, analytical-chemistry: tutorvista.com, Retrieved 19 April 2018

- Reactive Polymers Fundamentals and Applications: A Concise Guide to Industrial Polymers (Plastics Design Library), William Andrew Inc., 2nd edition, 2013, ISBN 978-1455731497

- Synthetic-rubber-production, rubber-chemical-compound: britannica.com, Retrieved 19 March 2018

- Hong Chua; Peter H. F. Yu & Chee K. Ma (March 1999). "Accumulation of biopolymers in activated sludge biomass". Applied Biochemistry and Biotechnology. Humana Press Inc. 78: 389–399. doi:10.1385/ABAB:78:1-3:389. ISSN 0273-2289. Retrieved 2009-11-24

- Types-of-synthetic-rubber: industrialrubbergoods.com, Retrieved 29 April 2018

- Khwaldia, Khaoula; Elmira Arab-Tehrany; Stephane Desobry (2010). "Biopolymer Coatings on Paper Packaging Materials". Comprehensive Reviews in Food Science and Food Safety. 9 (1): 82–91. doi:10.1111/j.1541-4337.2009.00095.x. Retrieved 9 Mar 2015

- Stemmelen, M.; Pessel, F.; Lapinte, V.; Caillol, S.; Habas, J.-P.; Robin, J.-J. (2011-06-01). "A fully biobased epoxy resin from vegetable oils: From the synthesis of the precursors by thiol-ene reaction to the study of the final material". Journal of Polymer Science Part A: Polymer Chemistry. 49 (11): 2434–2444. doi:10.1002/pola.24674. ISSN 1099-0518

# 2

# Polymers: Characteristics and Properties

The most basic characeteristics of a polymer are the identity of its monomer and its microstructure. Such properties influence the bulk physical properties of the polymer. This chapter delves into the characteristics and properties of polymers such as melting and boiling point, tensile strength, viscoelasticity, optical properties, crystallization of polymers, etc.

## Melting and Boiling Point

The melting point of a polymer does not occur over a sharp temperature range (1- 2 degrees C) as is observed for small organic molecules. If a polymer becomes a melt, there is usually a range of as much as 50 degrees C over which the viscosity of the polymer slowly changes from that of a solid to that of a liquid. Note that for a polymer to melt, a polymer must be a thermoplastic.

Using the term "melting point" is misleading, because polymers never "melt." The term "softening point" is sometimes used. A polymer does have a glass transition temperature (Tg) and a crystalline melt temperature (Tm.) The glass transition temperature refers to a point where there is a change in polymer molecular chain motion which has drastic effects on strength. Tg is sometimes called the "glass-rubber" transition. The crystalline melt temperature (Tm) is higher than Tg, and at Tm the crystalline domains of a polymer melt to become amorphous.

Data from Solomon's "Organic Chemistry"

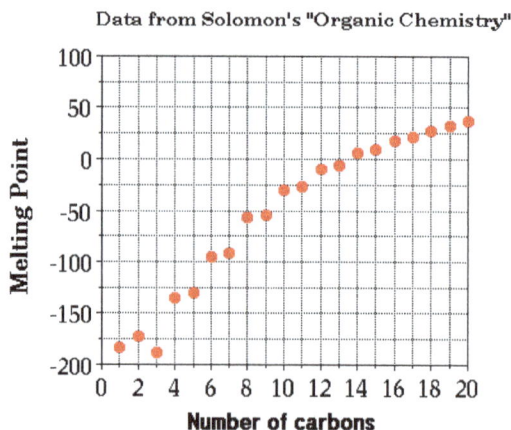

Melting pt. vs. # of carbons for straight chain alkane homologs

Boiling point- Polymers never boil.

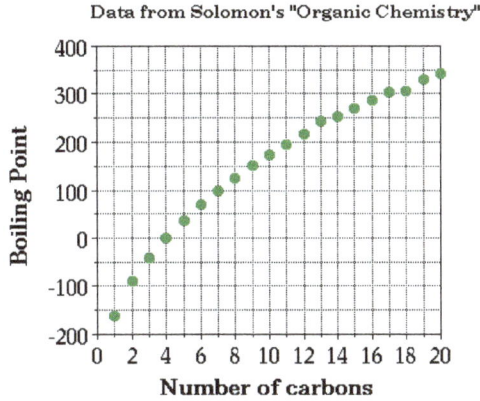

Data from Solomon's "Organic Chemistry"

Boiling pt. vs. # of carbons for straight chain alkane homologs

You may recall from Organic that as the number of carbons in the alkane homologous series increases, the boiling point increases asymptotically. We might chose to define the a polymer as a growing chain of sufficient length such that the mechanical properties are about 80% that of the apparent assomptotic limit. The number 80 is arbitrary. In the example below, chains of sufficient length to give a minimum performance for some mechanical property in the yellow-green zone are called polymers.

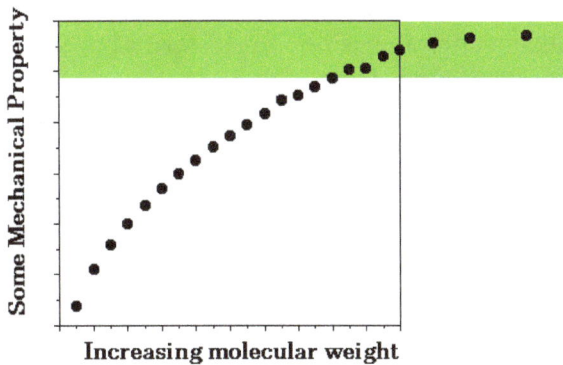

## Tensile Strength

The tensile strength of a material is the maximum amount of tensile stress that it can take before failure, such as breaking or permanent deformation. Tensile strength specifies the point when a material goes from elastic to plastic deformation. It is expressed as the minimum tensile stress (force per unit area) needed to split the material apart.

Tensile strength is calculated by dividing the load at break by the original minimum cross sectional area. The result is expressed in megapascals (MPA) and reported to three significant figures.

$$\text{tensile strength} = \frac{(\text{load at break})}{(\text{original width})(\text{original thickness})}$$

For example, if a metal rod one square inch in cross section can withstand a pulling force of 1,000 pounds but breaks if more force is applied, the metal has a tensile strength of 1,000 pounds per square inch. The tensile strength for structural steel is 400 megapascals (MPA) and for carbon steel is 841MPA. Tensile strength is different for different densities of steel.

There are three types of tensile strength:

1. Yield strength - The stress a material can withstand without permanent deformation.

2. Ultimate strength - The maximum stress a material can withstand.

3. Breaking strength - The stress coordinate on the stress-strain curve at the point of rupture.

Tensile strength is a limit state of tensile stress that leads to tensile failure in one of two manners:

1. Ductile failure - Yield as the first stage of failure, some hardening in the second stage and breakage after a possible "neck" formation.

2. Brittle failure - Sudden breaking in two or more pieces at a low stress state.

Tensile strength can be used in terms of either true stress or engineering stress.

Tensile strength testing for metal will determine how much a particular alloy will elongate before hitting ultimate tensile strength and how much load a particular piece of metal can accommodate before it loses structural integrity. Therefore, it is very important in material science. It is also vital for construction safety and personal safety, both during and after the building is completed.

Tensile strength, along with elastic modulus and corrosion resistance, is an important parameter of engineering materials that are used in structures and mechanical devices. It is specified for materials such as:

- Alloys

- Composite materials

- Ceramics

- Plastics

- Wood

# Tensile Properties of Polymers

The tensile properties of polymers are important for the design of plastic parts and the prediction of their performance under stress, particularly when used in structural applications. The simplest cases are homogeneous isotropic materials. For these materials, the mechanical response depends on only two constants, the Young modulus E, and the Poisson ratio $v$, whereas for anisotropic materials such as oriented-amorphous or oriented-crystalline polymers more constants are required to describe the mechanical response. Below, only isotropic materials will be considered.

At low strain ($< 1 \%$), the deformation of most polymeric materials is elastic, that is, the deformation is homogenous and after removal of the deforming load the material returns to its original size and shape. In this regime, the stress ($\sigma$) is directly proportional to the strain ($\varepsilon$), meaning, and the material obeys *Hooke's law*:

$$\sigma = E \, \varepsilon$$

where $E$ is the *Young's modulus* (also called *elastic modulus* or *tensile modulus*. It is the slope of the stress-strain curve, i.e., the ratio between an incremental increase in applied stress, $\Delta\sigma$, and an incremental deformation, $\Delta\varepsilon$. It is a measure for the stiffness of a material. The reciprocal of the Young's modulus is the *tensile compliance, D*, defined by

$$\varepsilon = D \, \sigma$$

In a *static stress-strain experiment*, a sample is pulled at a constant elongation rate and the stress, $\sigma$, is measured as function of strain, $\varepsilon$. A typical tensile test specimen is shown below.

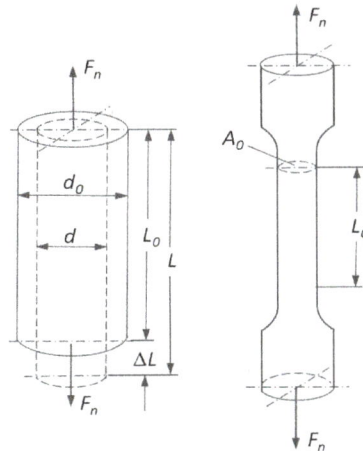

Tensile test specimen

In most cases, the stress-strain response of a tensile sample is reported in terms of nominal or engineering stress and strain. The *nominal or engineering tensile stress* is defined as the force divided by the initial (undeformed) cross-sectional area:

$$\sigma_e = F / A_o$$

and the *engineering tensile strain* or *Cauchy strain* is defined as the elongation, $\Delta L$, divided by the initial (undeformed) length of the specimen, $L_o$:

$$\varepsilon_e = \Delta L / L_o$$

Alternatively, the stress-strain response of a tensile sample may be reported in terms of *true stress* and *true strain*. The true stress is the ratio of the applied load to the actual cross-sectional area at a given elongation:

$$\sigma_t = F / A$$

If the volume of the test specimen is constant during deformation, it follows

$$L / L_o = A / A_o$$

Thus

$$\sigma_t = F / A_o \cdot (A_o / A) = \sigma_e \cdot (L_o / L)$$

The true strain (also called *Hencky strain*[2] or *logarithmic strain*) is the sum or integral of all incremental length changes, $dL$, divided by the actual length $L$:

$$\varepsilon_T = \int dL / L = \ln(L / L_o)$$

The relationship between true strain and engineering strain is simply

$$\varepsilon_T = \ln(L / L_o) = \ln[(L_o + \Delta L) / L_o] = \ln(1 + \varepsilon_e)$$

These relations are only valid up to the onset of necking. Beyond this point, strong necking will result in a change in volume.

For small elongations, true stress and engineering stress are essentially the same, whereas for large elongations the true strain and stress are preferred.

## Viscoelasticity

Viscoelasticity is the property of materials that exhibit both viscous and elastic characteristics when undergoing deformation. A viscous material exhibits time-dependent behavior when a stress is applied while under constant stress and deforms at a constant rate, and when the load is removed, the material has 'forgotten' its original configuration, remaining in the deformed state. On the other hand, an elastic material deforms instantaneously when stretched and 'remembers' its original configuration, returning instantaneously to its original state once the stress is removed. Viscoelastic materials have elements of both of these properties and, as such, exhibit time-dependent strain showing a 'fading memory'. Such a behavior may be linear (stress and strain are proportional) or nonlinear. Whereas elasticity is usually the result of bond stretching along crystallographic planes in an ordered solid, viscoelasticity is the result of the diffusion of atoms or molecules inside an amorphous material.

The typical response of a viscoelastic material is as sketched in Figure. The following will be noted:

1.  The loading and unloading curves do not coincide, Figure a, but form a hysteresis loop.

2.  There is a dependence on the rate of straining $d\varepsilon / dt$ , Figure b; the faster the stretching, the larger the stress required.

3.  There may or may not be some permanent deformation upon complete unloading, figure.

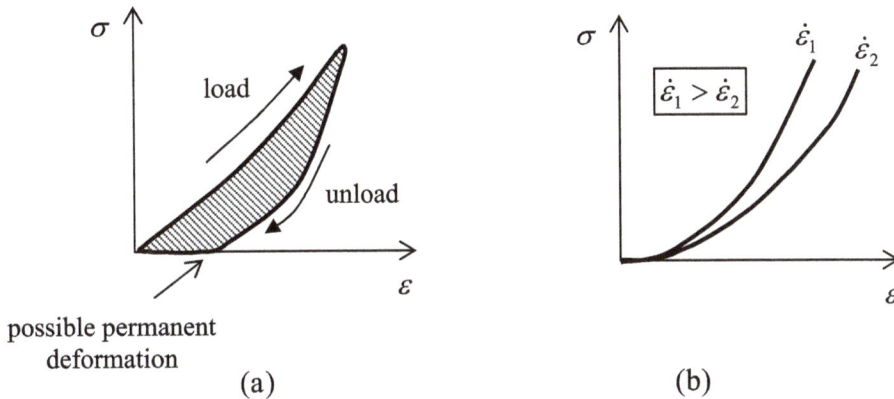

possible permanent
deformation
(a)                                              (b)

Figure: Response of a Viscoelastic material in the Tension test;
(a) loading and unloading with possible permanent deformation
(non-zero strain at zero stress), (b) different rates of stretching

The effect of rate of stretching shows that the viscoelastic material depends on time. This contrasts with the elastic material, whose constitutive equation is independent of time, for example it makes no difference whether an elastic material is loaded to some given stress level for one second or one day, or loaded slowly or quickly; the resulting strain will be the same.

## Linear Viscoelasticity

Linear viscoelastic materials are those for which there is a linear relationship between stress and strain (at any given time), $\sigma \propto \varepsilon$ . As mentioned before, this requires also that the strains are small, so that the engineering strain measure can be used (since the exact strain is inherently non-linear).

Strain-time curves for a linear viscoelastic material subjected to various constant stresses are shown in Figure. At any given time, say $t_1$ , the strain is proportional to stress, so that the strain there due to $3\sigma_o$ is three times the strain due to $\sigma_o$ .

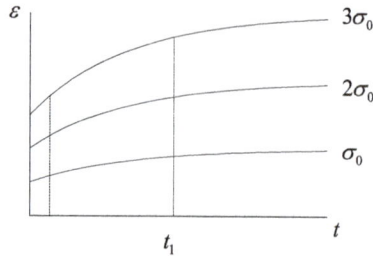

Figure: Strain as a function of time at different loads

Linear viscoelasticity is a reasonable approximation to the time-dependent behaviour of metals and ceramics at relatively low temperatures and under relatively low stress. However, its most widespread application is in the modelling of polymers.

## Testing of Viscoelastic Materials

A number of other tests which are especially useful for the characterisation of viscoelastic materials have been developed.

## The Creep and Recovery Test

The creep-recovery test involves loading a material at constant stress, holding that stress for some length of time and then removing the load. The response of a typical viscoelastic material to this test is show in Figure.

First there is an instantaneous elastic straining, followed by an ever-increasing strain over time known as creep strain. The creep strain usually increases with an ever de-creasing strain rate so that eventually a more-or-less constant-strain steady state is reached, but many materials often do not reach such a noticeable steady-state, even after a very long time.

When unloaded, the elastic strain is recovered immediately. There is then anelastic recovery – strain recovered over time; this anelastic strain is usually very small for metals, but may be significant in polymeric materials.  A permanent strain may then be left in the material.

A test which focuses on the loading phase only is simply called the creep test.

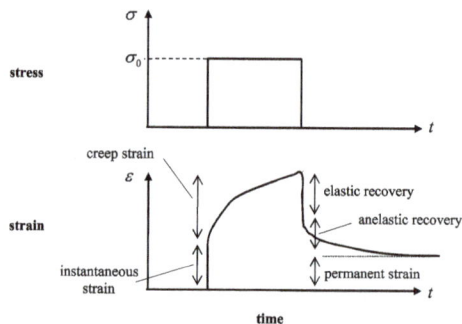

Figure: Strain response to the creep-recovery test

## Stress Relaxation Test

The stress relaxation test involves straining a material at constant strain and then holding that strain, Figure. The stress required to hold the viscoelastic material at the constant strain will be found to decrease over time. This phenomenon is called stress relaxation; it is due to a re-arrangement of the material on the molecular or micro-scale.

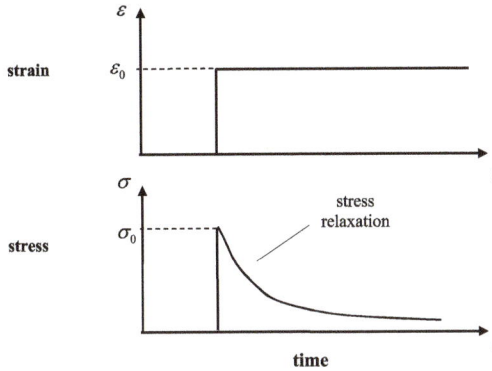

Figure: Stress response to the stress-relaxation test

## The Cyclic Test

The cyclic test involves a repeating pattern of loading-unloading, Figure. It can be strain-controlled (with the resulting stress observed), as in Figure, or stress-controlled (with the resulting strain observed). The results of a cyclic test can be quite complex, due to the creep, stress-relaxation and permanent deformations.

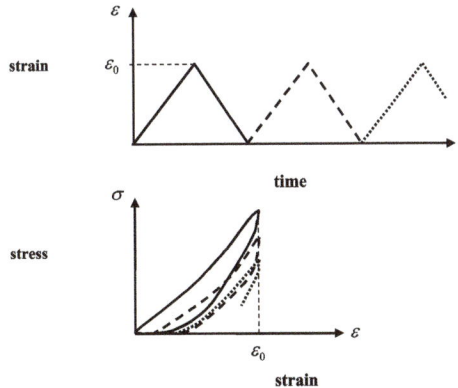

Figure: Typical stress response to the cyclic test

Nonlinear viscoelasticity is when the function is not separable. It usually happens when the deformations are large or if the material changes its properties under deformations.

An anelastic material is a special case of a viscoelastic material: an anelastic material will fully recover to its original state on the removal of load.

## Dynamic Modulus

Viscoelasticity is studied using dynamic mechanical analysis, applying a small oscillatory stress and measuring the resulting strain.

- Purely elastic materials have stress and strain in phase, so that the response of one caused by the other is immediate.

- In purely viscous materials, strain lags stress by a 90 degree phase lag.

- Viscoelastic materials exhibit behavior somewhere in the middle of these two types of material, exhibiting some lag in strain.

Complex Dynamic modulus G can be used to represent the relations between the oscillating stress and strain:

$$G = G' + iG''$$

where $i^2 = -1$; $G'$ is the *storage modulus* and $G''$ is the *loss modulus*:

$$G' = \frac{\sigma_0}{\varepsilon_0} \cos \delta$$

$$G'' = \frac{\sigma_0}{\varepsilon_0} \sin \delta$$

Where $\sigma_0$ and $\varepsilon_0$ are the amplitudes of stress and strain respectively, and $\delta$ is the phase shift between them.

## Constitutive Models of Linear Viscoelasticity

Viscoelastic materials, such as amorphous polymers, semicrystalline polymers, biopolymers and even the living tissue and cells, can be modeled in order to determine their stress and strain or force and displacement interactions as well as their temporal dependencies. These models, which include the Maxwell model, the Kelvin–Voigt model, the Standard Linear Solid model, and the Burgers model, are used to predict a material's response under different loading conditions. Viscoelastic behavior has elastic and viscous components modeled as linear combinations of springs and dashpots, respectively. Each model differs in the arrangement of these elements, and all of these viscoelastic models can be equivalently modeled as electrical circuits. In an equivalent electrical circuit, stress is represented by voltage, and strain rate by current. The elastic modulus of a spring is analogous to a circuit's *capacitance* (it stores energy) and the viscosity of a dashpot to a circuit's *resistance* (it dissipates energy).

The elastic components, as previously mentioned, can be modeled as springs of elastic constant E, given the formula:

$$\sigma = E\varepsilon$$

where σ is the stress, E is the elastic modulus of the material, and ε is the strain that occurs under the given stress, similar to Hooke's Law.

The viscous components can be modeled as dashpots such that the stress–strain rate relationship can be given as,

$$\sigma = \eta \frac{d\varepsilon}{dt}$$

where σ is the stress, η is the viscosity of the material, and dε/dt is the time derivative of strain.

The relationship between stress and strain can be simplified for specific stress rates. For high stress states/short time periods, the time derivative components of the stress–strain relationship dominate. A dashpot resists changes in length, and in a high stress state it can be approximated as a rigid rod. Since a rigid rod cannot be stretched past its original length, no strain is added to the system.

Conversely, for low stress states/longer time periods, the time derivative components are negligible and the dashpot can be effectively removed from the system - an "open" circuit. As a result, only the spring connected in parallel to the dashpot will contribute to the total strain in the system.

## Maxwell Model

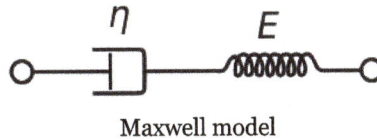

Maxwell model

The Maxwell model can be represented by a purely viscous damper and a purely elastic spring connected in series, as shown in the diagram. The model can be represented by the following equation:

$$\sigma + \frac{\eta}{E}\dot{\sigma} = \eta\dot{\varepsilon}$$

Under this model, if the material is put under a constant strain, the stresses gradually relax. When a material is put under a constant stress, the strain has two components. First, an elastic component occurs instantaneously, corresponding to the spring, and relaxes immediately upon release of the stress. The second is a viscous component that grows with time as long as the stress is applied. The Maxwell model predicts that stress decays exponentially with time, which is accurate for most polymers. One limitation of this model is that it does not predict creep accurately. The Maxwell model for creep or constant-stress conditions postulates that strain will increase linearly with time. However, polymers for the most part show the strain rate to be decreasing with time.

Applications to soft solids: thermoplastic polymers in the vicinity of their melting temperature, fresh concrete (neglecting its aging), numerous metals at a temperature close to their melting point.

## Kelvin–voigt Model

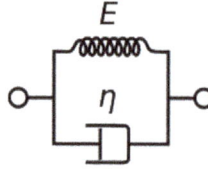

Schematic representation of Kelvin–Voigt model

The Kelvin–Voigt model, also known as the Voigt model, consists of a Newtonian damper and Hookean elastic spring connected in parallel, as shown in the picture. It is used to explain the creep behaviour of polymers.

The constitutive relation is expressed as a linear first-order differential equation:

$$\sigma = E\varepsilon + \eta\dot{\varepsilon}$$

This model represents a solid undergoing reversible, viscoelastic strain. Upon application of a constant stress, the material deforms at a decreasing rate, asymptotically approaching the steady-state strain. When the stress is released, the material gradually relaxes to its undeformed state. At constant stress (creep), the Model is quite realistic as it predicts strain to tend to $\sigma/E$ as time continues to infinity. Similar to the Maxwell model, the Kelvin–Voigt model also has limitations. The model is extremely good with modelling creep in materials, but with regards to relaxation the model is much less accurate.

Applications: organic polymers, rubber, wood when the load is not too high.

## Standard Linear Solid Model

The standard linear solid model, also known as the Zener model, consists of two springs and a dashpot. It is the simplest model that describes both the creep and stress relaxation behaviors of a viscoelastic material properly. For this model, the governing constitutive relations are:

Maxwell representation

Kelvin representation

$$\sigma + \frac{\eta}{E_2}\dot{\sigma} = E_1\varepsilon + \frac{\eta(E_1 + E_2)}{E_2}\dot{\varepsilon}$$

$$\sigma + \frac{\eta}{E_1 + E_2}\dot{\sigma} = \frac{E_1 E_2}{E_1 + E_2}\varepsilon + \frac{E_1\eta}{E_1 + E_2}\dot{\varepsilon}$$

Under a constant stress, the modeled material will instantaneously deform to some strain, which is the instantaneous elastic portion of the strain. After that it will continue to deform and asymptotically approach a steady-state strain, which is the retarded elastic portion of the strain. Although the Standard Linear Solid Model is more accurate than the Maxwell and Kelvin–Voigt models in predicting material responses, mathematically it returns inaccurate results for strain under specific loading conditions.

## Burgers Model

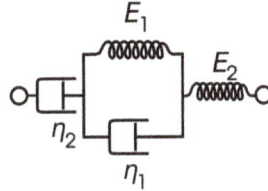

Schematic representation of Burgers model

The Burgers model combines the Maxwell and Kelvin–Voigt models in series. The constitutive relation is expressed as follows:

$$\sigma + \left( \frac{\eta_1}{E_1} + \frac{\eta_2}{E_1} + \frac{\eta_2}{E_2} \right) \dot{\sigma} + \frac{\eta_1 \eta_2}{E_1 E_2} \ddot{\sigma} = \eta_2 \dot{\varepsilon} + \frac{\eta_1 \eta_2}{E_1} \ddot{\varepsilon}$$

This model incorporates viscous flow into the standard linear solid model, giving a linearly increasing asymptote for strain under fixed loading conditions.

## Generalized Maxwell Model

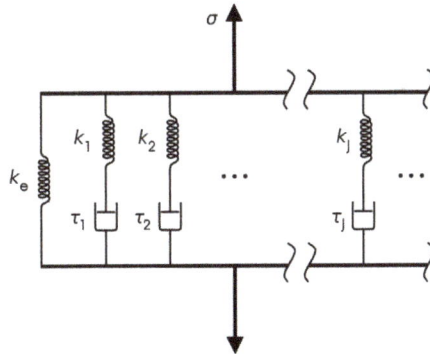

Schematic of Maxwell-Wiechert Model

The Generalized Maxwell model, also known as the Wiechert model, is the most general form of the linear model for viscoelasticity. It takes into account that the relaxation does not occur at a single time, but at a distribution of times. Due to molecular segments of different lengths with shorter ones contributing less than longer ones, there is a varying time distribution. The Wiechert model shows this by having as many spring–dashpot Maxwell elements as are necessary to accurately represent the distribution.

The figure on the right shows the generalised Wiechert model. Applications: metals and alloys at temperatures lower than one quarter of their absolute melting temperature (expressed in K).

## Prony Series

In a one-dimensional relaxation test, the material is subjected to a sudden strain that is kept constant over the duration of the test, and the stress is measured over time. The initial stress is due to the elastic response of the material. Then, the stress relaxes over time due to the viscous effects in the material. Typically, either a tensile, compressive, bulk compression, or shear strain is applied. The resulting stress vs. time data can be fitted with a number of equations, called models. Only the notation changes depending of the type of strain applied: tensile-compressive relaxation is denoted $E$, shear is denoted $G$, bulk is denoted $K$. The Prony series for the shear relaxation is

$$G(t) = G_\infty + \Sigma_{i=1}^N G_i \exp(-t / \tau_i)$$

where $G_\infty$ is the long term modulus once the material is totally relaxed, $\tau_i$ are the relaxation times (not to be confused with $\tau_i$ in the diagram); the higher their values, the longer it takes for the stress to relax. The data is fitted with the equation by using a minimization algorithm that adjust the parameters ($G_\infty, G_i, \tau_i$) to minimize the error between the predicted and data values.

An alternative form is obtained noting that the elastic modulus is related to the long term modulus by

$$G(t = 0) = G_0 = G_\infty + \Sigma_{i=1}^N G_i$$

Therefore,

$$G(t) = G_0 - \Sigma_{i=1}^N G_i [1 - \exp(-t / \tau_i)]$$

This form is convenient when the elastic shear modulus $G_0$ is obtained from data independent from the relaxation data, and/or for computer implementation, when it is desired to specify the elastic properties separately from the viscous properties, as in.

A creep experiment is usually easier to perform than a relaxation one, so most data is available as (creep) compliance vs. time. Unfortunately, there is no known closed form for the (creep) compliance in terms of the coefficient of the Prony series. So, if one has creep data, it is not easy to get the coefficients of the (relaxation) Prony series, which are needed for example in. An expedient way to obtain these coefficients is the following. First, fit the creep data with a model that has closed form solutions in both compliance and relaxation; for example the Maxwell-Kelvin model in or the Standard Solid Model in (section). Once the parameters of the creep model are known, produce relaxation pseudo-data with the conjugate relaxation model for the same times of the original data. Finally, fit the pseudo data with the Prony series.

## Effect of Temperature on Viscoelastic Behavior

The secondary bonds of a polymer constantly break and reform due to thermal motion. Application of a stress favors some conformations over others, so the molecules of the polymer will gradually "flow" into the favored conformations over time. Because thermal motion is one factor contributing to the deformation of polymers, viscoelastic properties change with increasing or decreasing temperature. In most cases, the creep modulus, defined as the ratio of applied stress to the time-dependent strain, decreases with increasing temperature. Generally speaking, an increase in temperature correlates to a logarithmic decrease in the time required to iMPArt equal strain under a constant stress. In other words, it takes less work to stretch a viscoelastic material an equal distance at a higher temperature than it does at a lower temperature.

Extreme cold temperatures can cause viscoelastic materials to change to the glass phase and become brittle. For example, exposure of pressure sensitive adhesives to extreme cold (dry ice, freeze spray, etc.) causes them to lose their tack, resulting in debonding.

## Optical Properties

The optical properties of polymers, including haze, gloss, and transparency, are critical to many applications of commercial plastics. The *transmittance* of a material is defined as the ratio of light intensity passing through the material to the intensity of light received by the specimen. It is determined by *reflection, absorption* and *scattering*. If both absorption and scattering are negligible, the material is called *transparent*. In contrast, an *opaque* material has practically zero transmittance due to its high scattering power whereas a materials with negligible absorption but with appreciable transmittance but lower than 90% is called *translucent*.

Gloss is an optical property which describes how well a surface reflects light in specular direction. It is of great practical importance for many applications. The *reflectivity* (or reflectance) is defined as the fraction of incident light intensity reflected at the interface (surface effect). Only highly polished metal mirrors have nearly total reflectance. If the reflectance is almost zero, the surface appears totally matt. A material with reflectance in between these two extremes is called glossy or shiny. The gloss is responsible for the lustrous appearance of plastic films.

The fraction of the light intensity that neither enters the material nor follows the pass of mirror reflection is dispersed or scattered by *diffraction*. The lustrous appearance of a plastic is determined by both the reflection and diffraction, meaning the shininess of a surface depends on both the specular reflection and the diffuse light of the surrounding surface area, which is often called the *contrast gloss* or *luster*.

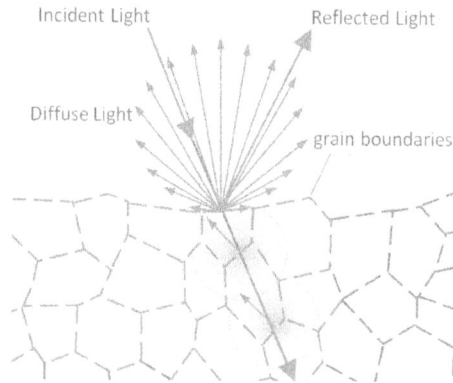

The root cause of opacity in polymers is light scattering and absorption. The more light is scattered or absorbed the more the material becomes opaque. Light scattering is a purely physical phenomenon: light hitting insulating matter induces dipole oscillations in the material. Each induced dipole then acts as a secondary source of light and emits photons in all directions, the so called scattered light. This, in turn, reduces the intensity of directly transmitted light. Pure light scattering will not cause any loss of radiation energy. However in the case of colored or filled plastic, a portion of the light is absorbed by the material, that is, radiation energy is transformed into other forms of energy like induced molecule motion (thermal energy). Assuming a linear absorption mechanism the decrease in radiation intensity as a function of penetration depth can be described by the Lambert absorption rule

$$I_T(z) = I_o \cdot \exp(-\alpha_A \cdot z)$$

where $z$ is the penetration depth of the light and $\alpha_A = 4 \pi \kappa / \lambda$ is the wave length dependent absorption constant. The Lambert absorption rule states that within every differential layer of material, dz, the same fraction of radiation is absorbed which leads to an exponential decrease of transmitted radiation.

Light scattering in clear polymers is caused by optical inhomogeneity such as phase boundaries, pores and inclusions, to be more specific, light scattering in clear plastics arises from fluctuations in refractive index due to fluctuations in composition and/or packing density (volume effect) whereas a defect-free crystalline solid provides no scattering centers for incoming lightwaves. However, polymers crystals have usually many defects like phase boundaries between crystals and amorphous regions. The intensity loss due to scattering caused by inhomogeneous structures in the material can also be described with the exponential rule of Lambert. In polymeric materials, scattering of radiation like NIR and IR can be caused by the macromolecules themselves, by crystalline-amorphous phase boundaries as well as by filler particles, fibers and pigments. The type of scattering depends on the ratio of wave length $\lambda$ to the size of scattering structure $s$.

Following cases exist:

    I.   $s " \lambda$ : Diffraction on microscopic structures,

II.  $s \geq \lambda$: Mie scattering on small scattering structures,

III.  $s \text{``} \lambda$: Rayleigh and Raman scattering.

In the case of diffraction on microscopic structures like phase boundaries, scattering is independent of the wave length. This type of diffraction often dominates in thin colorless plastic films.

The drop in radiation intensity caused by the overall scattering can by described by the exponential rule of Lambert:

$$I_T(z) = I_o \cdot \exp(-\alpha_S \cdot z)$$

where $\alpha_S$ is the scattering coefficient of the medium. For *spherical* scattering structures smaller than the wavelength (Rayleigh scattering), the scattering constant is proportional to the number $N$ and the volume $V$ of the scattering structures as well as the wavelength of the radiation $\lambda$,

$$\alpha_S = N \cdot V / \lambda^4$$

In the case of large *spherical* scattering structures (diffraction on microscopic structures), the scattering constant is proportional to the sum of the cross-sectional areas of the scattering elements:

$$\alpha_S \propto \pi \sum_i \rho_i^2$$

The relationship above are only valid for spherical scattering structures like aerogels. For crystalline polymers, the mathematical description of the scattering constants is considerably more complex.

Despite many defects, thin plastic films sometimes appear perfectly clear. The explanation is simple; only when the heterogeneities are in the range of visible light, significant reflection, refraction, and scattering can be expected and the polymer film loses clarity. If, however, these regions are smaller than the wavelength of the light, the polymer appears clear.

The size and degree of crystallinity depends on many factors such as polymerization and heat-treatment conditions, composition, structure and size of the polymer chains including polydispersity and chain branching. The clarity or transmittance of a polymer usually increases with decreasing crystallinity, refractive index, compressibility and intermolecular interaction (cohesive energy density).

The degree of scattering also depends on the surface morphology. Often a large portion of the incident light is scattered at the polymer surface as it is the case for polymer films. Scattering is caused by surface irregularities (roughness) and imperfections such as scratches (surface deffects). Surface roughness is often the most important factor. For example, it has been shown that HDPE lamellae can form large rodlike superstructures which produce rough surfaces. These surface irregularities can scatter a large portion of the incident light and cause haziness whereas polymers with less oriented lamellar structures have reduced surface roughness (smaller domains) and thus are less hazy.

For a perfectly smooth surface the specular reflectance, $R_s = I_R / I_o$, for unpolarized light of intensity $I_o$ can be calculated with the Fresnel equation:

$$R_s = \frac{I_R}{I_o} = \frac{1}{2}\left|\left(\frac{\cos\theta - \sqrt{n^2 - \sin^2\theta}}{\cos\theta + \sqrt{n^2 - \sin^2\theta}}\right)^2 + \left(\frac{n^2\cos\theta - \sqrt{n^2 - \sin^2\theta}}{n^2\cos\theta + \sqrt{n^2 - \sin^2\theta}}\right)^2\right|$$

where $n$ is the refractive index of the surface and $\theta$ is the angle of incidence. Due to surface irregularities and other imperfections, a rough surface will scatter some light in non-specular directions. This reduces the measured gloss. At a given angle $\theta$ and wavelength $\lambda$, there will be a transition from smooth to rough surface. As has been shown by Bennett and Porteus (1961) the effect of surface roughness on the reflectance can be described by the equation

$$R_r = R_s \cdot \exp\{- [(4n\sigma / \lambda) \cos \theta]^2\}$$

where $R_r$ is the specular reflectance of the rough surface and $\sigma$ is the root mean square surface roughness in the units of wave length of the light.

## Colour Measurement

Most polymers are typically grey or yellow coloured. White fillers or other optically active additives colour additives can help to make the appearance more suited to application.

## Yellowness Index

Yellowness Index is a number calculated from spectrophotometric data that describes the change in color of a test sample from clear or white to yellow. Intertek uses the yellowness index test to evaluate color changes in a material caused by real or simulated outdoor exposure.

## Haze

Haze is an optical effect caused by light scattering within a transparent polymer resulting in a cloudy or milky appearance. Our polymer scientists use spectrophotometer techniques to investigate a range of issues which result in haze as a symptom such as weathering, during product and process development.

## Birefringence

The refractive index varies with orientation of the polymer chains on a molecular level; our polymer scientists use polarized light to achieve birefringence effects which can be used to quantify the stress in a transparent plastic.

# Crystallization of Polymers

Crystalline polymers have a less perfect structure than crystals formed from low molecular weight compounds. A common basic structure are lamellae that consist of layers of folded chains as illustrated below. The loops of the lamellae can be loose and irregular or thigh and regular. The thickness of a typical crystallite is in the range of 10 - 20 nm. In the absence of a thermal gradient, the lamellae grow radially in all directions, resulting in spherical crystalline regions the so called spherulites. Usually, polymers can only produce partially crystalline structures, i.e. they are semi crystalline because polymers do not have a uniform molecular weight.

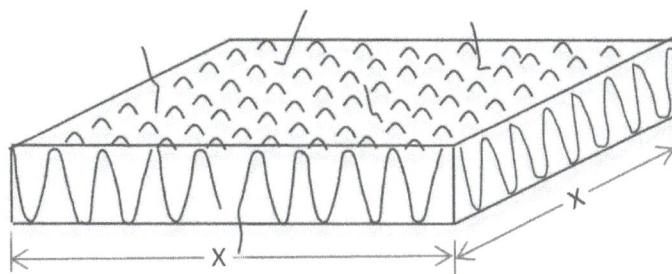

Crystallinity is usually induced by cooling a melt or a dilute solution below its melting point. The later can result in the growth of single crystals. Crystallization can also be induced by stretching a polymer. In this case, crystallization is caused by molecular orientation in the stretch direction. If the temperature is above the brittle point (preferentially above the $Tg$) and the polymer is stretched, the randomly coiled and entangled chains begin to disentangle, unfold, and straighten. This method is called *strain-induced crystallization*. It occurs when polymers are stretched beyond its yield point. One usually observes a noticeable increase in modulus due to the formation of crystals that act as a physica reinforcements similar to fillers. Thus when strain induced crystallization occurs, the stress increases as well.

The size and structure of the crystals and the degree of crystallinity depend on the type and structure of the polymer, and on the growth conditions. Narrow molecular weight, linear polymer chains, and high molecular weight increase the crystallinity. Crystallinity is also affected by extrinsic factors, like crystallization temperature, cooling rate, and in the case of strain-induced crystallization, by the stretch ratio, strain rate, and by the forming process of the polymer film or fiber.

Nucleating agents such as organic salts, small filler particles and ionomers also affect the crystallization. They act as seeds and can increase the crystallization rate.

The degree of crystallinity also depends on the tacticity of the polymer. The greater the order in a macromolecule the greater the likelihood of the molecule to undergo crystallization. For example, isotactic polypropylene is usually more crystalline than syndiotactic polypropylene, and atactic polypropylene is considered uncrystallizable since the structure of the polymer chain lacks any regularity. In fact, most atactic polymers do not crystallize.

Strong intermolecular forces and a stiff chain backbone favor the formation of crystals because the molecules prefer an ordered arrangement with maximum packing density to maximize the number of secondary bonds. Thus the molecules tend to cooperatively organize and develop a crystalline structure. A good example is Kevlar which has a high degree of crystallinity. The polar amide groups in the backbone are strongly attracted to each other and form strong hydrogen bonds. This raises the glass transiton temperature and the melting point. The high crystallinity and strong intermolecular interactions also greatly increases the mechanical strength. In fact, Kevlar fibers are some of the strongest plastic fibers on the market.

Bulky side groups have the opposite effect on crystallinity. With increasing size of the side groups it becomes progressively more difficult for the polymer to fold and align itself along the crystal growth direction. Thus bulky side groups and branching reduce the ability and likelihood of a polymer to crystallize. For example branched polyethylene has a low dregree of crystallinity, even though polyethylene itself easily crystallizes. Similarly, most network polymers do not crystallize because the polymer subchains do not have the freedom to move.

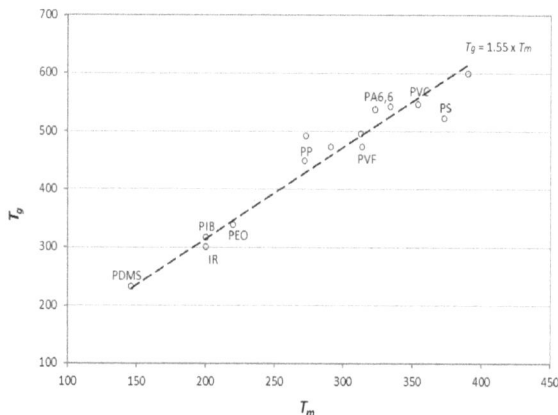

Glass transition temperature versus melting point

Crystalline polymers are characterized by a melting point $Tm$ and amorphous polymers are characterized by a glass transition temperature $Tg$. For crystalline polymers, the relationship between $Tm$ and $Tg$ has been described by Boyer as follows:

$Tg / Tm \approx 1 / 2 \rightarrow$ symmetrical polymers

$Tg / Tm \approx 2 / 3 \rightarrow$ unsymmetrical polymers

It was found that many polymers with a $Tg / Tm$ ratio below 1/2 are highly symmetrical and consist of small repeating units of one or two main-chain atoms each carrying only single atom substituents. Examples are poly(methylene oxide), polyethylene, and polyacetal. These polymers are markedly crystalline. Polymers with $Tg$ / $Tm$ ratios above 2/3 are usually unsymmetrical. They can be highly crystalline if they have long sequences of methylene groups or are highly stereo-regular. The majority of the polymers, however, have $Tg$ / $Tm$ ratios between 0.5 and 0.75 with a maximum number around 2/3 (see figure above); both symmetrical and unsymmetrical polymers belong to this group.

## References

- Tensile-strength-1072: corrosionpedia.com, Retrieved 15 May 2018

- Biswas, Abhijit; Manivannan, M.; Srinivasan, Mandyam A. (2015). "Multiscale Layered Biomechanical Model of the Pacinian Corpuscle". IEEE Transactions on Haptics. 8 (1): 31–42. doi:10.1109/TOH.2014.2369416. PMID 25398182

- Mechanical-Properties, Polymer-physics: polymerdatabase.com: Retrieved 22 April 2018

- Viscoelasticity, chemistry: sciencedirect.com, Retrieved 31 March 2018

- Optical-properties, polymers: intertek.com, Retrieved 28 June 2018

# Inorganic Polymers

The polymer which has a skeletal structure but does not have carbon atoms in its backbone is referred to as inorganic polymer. The chapter closely examines some inorganic polymers such as polyphosphazene, ionomer, geopolymer and conductive polymers to provide an extensive understanding of the subject.

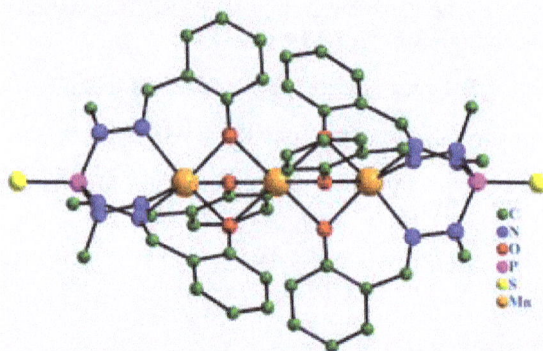

An inorganic polymer is defined as a giant molecule linked by covalent bonds but with an absence or near-absence of hydrocarbon units in the main molecular backbone; these may be included as pendant side chains. Carbon fibers, graphite, and so forth are considered inorganic polymers. Much of inorganic chemistry is the chemistry of high polymers.

For compounds that do not melt or dissolve without chemical change, both the absence of an equilibrium vapor pressure and the observation of a dissociation pressure resulting from de-polymerization bring them into the framework of the definition.

## Properties

Some special characteristics of many inorganic polymers are a higher Young's modulus and a lower failure strain compared with organic polymers. Relatively few inorganic polymers dissolve in the true sense, or alternatively, if they swell, few can revert. Crystallinity and high glass transition temperatures are also much more common than in organic polymers. In highly cross-linked inorganic polymers, stress relaxation frequently involves bond interchange.

The properties of inorganic polymers require a different technology from that of their organic counterparts. Such technology is either completely new (such as reconstructive processing — the spinning of an inorganic compound on an organic support or binder subsequently removed by oxidation/volatilization), or it has been adapted from other fields, for example, glass technology. Thus, reconstructed vermiculite can give flexible sheets. Yarn, paper, woven cloth, and even textiles can be made from alumina and zirconia fibers by the spinning/volatilization process. A mica-forming glass ceramic is resistant to thermal and mechanical shock and can be worked with conventional metalworking tools.

## Classification

Inorganic polymers can be classified in a number of ways. Some are based on the composition of the backbone, such as the silicones (Si-O), the phosphazenes (P-N), and polymeric sulfur (S-S). Others are based on their connectivity, that is, the number of network bonds linking the repeating unit into the network. Thus, the silicones based on $R_2SiO$, the phosphazenes based on $NPX_2$, and polymeric sulfur each have a connection of two, while boric oxide based on $B_2O_3$ has a connectivity of three, and amorphous silica based on $SiO_2$ has a connection of four.

## Types

The number of inorganic polymers is very large. Sulfur, selenium, and tellurium all form high polymers. Polymers of sulfur are usually elas-tomeric, and those of selenium and tellurium are generally crystalline. In the melt at 220 °C, the molecular weight of the sulfur polymer is about 12,000,000 and that of selenium about 800,000.

## Silicones

Perhaps best known of all the synthetic polymers based on inorganic molecular structures are the silicones, which are derived from the basic units.

## Chalcogenide Glasses

These are amorphous cross-linked polymers with a connectivity of three. Probably the best known is arsenic sulfide, $(As_2S_3)n$, which can be used for infrared transparent windows. Threshold and memory switching are also interesting properties of these glasses. Ultraphosphate glasses resemble glassy organic plastics and can be processed by the same methods, such as extrusion and injection molding. They are used for antifouling surfaces for marine applications and in the manufacture of no misting spectacle lenses.

## Graphite

This is a well-known two-dimensional polymer with lubricating and electrical properties. Intercalation compounds of graphite can have super-metallic anisotropic properties.

## Boron Polymers

Structurally related to graphite is hexagonal boron nitride (BN). Like graphite, it has lubricating properties, reflecting the relationship between molecular structure and physical properties, but unlike graphite, it is an electrical insulator. Molybdenum di-sulfide, $(MoS_2)n$, with a similar and related structure, is also a solid lubricant. Both graphite and hexagonal boron nitride can be readily machined. Outstanding properties of the latter include high thermal and chemical stability and good dielectric properties. Crucibles and such items as nuts and bolts can be made from this material.

Borate glasses with comparatively low softening points are used as solder and sealing glasses and can be prepared by fusing mixtures of metal oxides with boric oxide, $(B_2O_3)n$.

## Silicate Polymers

The silicates, both crystalline and amorphous, supply a very large number of inorganic polymers. Examples include the naturally occurring fiberlike asbestos and sheetlike mica. The industrially important water-soluble alkali metal silicates can give highly viscous polymeric solutions. Borosilicate glasses form another important group of silicate polymers. The Pyrex type is well known for its resistance to thermal shock; the leached Vycor type is porous and can be used for filtering bacteria and viruses. Asbestos occurs as ladder polymers, of which crocidolite is the most important, and as layer polymer, exemplified by chrysotile. The zeolites, many of which have been found naturally or have been synthesized, are three-dimensional network polymers. Their uses as molecular sieves are well known.

## Other Polymers

Silicon nitride, $(Si_3N_4)n$, is another macromol-ecule with interesting properties. Prepared by heating of silicon powder in an atmosphere of nitrogen (nitridation) at above 1200 °C, the product is a material that can be machined readily and whose good thermal shock resistance and creep resistance at high temperatures, which is further improved by admixture of another inorganic macromolecule silicon carbide, make it useful for applications in gas turbine, diesel engines, thermocouple sheaths, and a variety of com-ponents.

Allotropic forms of carbon boron nitride are diamond and cubic boron nitride, both preparable by high-temperature and high-pressure syntheses and characterized by extreme hardness, which make them useful industrially in cutting and grinding tools.

## Importance

Inorganic polymeric materials are growing in importance as a result of a combination of two major factors: the depletion of the world's fossil fuel reserves (the basis of the petrochemical industry) and the ever-increasing demands of modern technology, coupled with environmental and health regulations, such as flame retardancy and non-flammability.

## Geopolymer

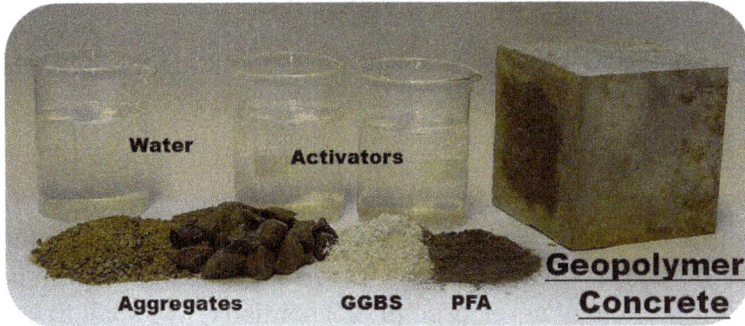

Geopolymers are chains or networks of mineral molecules linked with co-valent bonds. They have following basic characteristics:

a) *Nature of the hardened material:*

- X-ray amorphous at ambient and medium temperatures.

- X-ray crystalline at temperatures > 500°C.

b) *Synthesis Routes:*

- Alkaline medium (Na, K, Ca) hydroxides and alkali-silicates yielding poly(silicates) – poly(siloxo) type or poly(silico-aluminates) – poly(sialate) type.

- Acidic medium (Phosphoric acid) yielding poly(phospho-siloxo) and poly(alumino-phospho) types.

As an example, one of the geopolymeric precursors, MK-750 (metakaolin) with its alumoxyl group –Si-O-Al=O, reacts in both systems, alkaline and acidic. Same for siloxo-based and organo-siloxo-based geopolymeric species that also react in both alkaline and acidic medium.

## Geopolymer Terminology

In the late 1970's, Joseph Davidovits, the inventor and developer of geopolymerization, coined the term "geopolymer" to classify the newly discovered geosynthesis that produces inorganic polymeric materials now used for a number of industrial applications. He also set a logical scientific terminology based on different chemical units, essentially for silicate and aluminosilicate materials, classified according to the Si:Al atomic ratio:

$$Si:Al = 0, siloxo$$

$$Si:Al = 1, sialate\ (acronym\ for\ silicon\text{-}oxo\text{-}aluminate\ of\ Na,\ K,\ Ca,\ Li)$$

$$Si:Al = 2, sialate\text{-}siloxo$$

$$Si:Al = 3, sialate\text{-}disiloxo$$

$$Si:Al > 3, sialate\ link.$$

This terminology was presented to the scientific community at a IUPAC conference in 1976.

The geopolymeric '*sialate*' term proceeds from the same scientific logic (it is the acronym of silicon-oxo-aluminate), in contrast with the organic molecule 'sialic acid' that was derived from an ancient Greek word meaning 'saliva', with no scientific association. In fact, for our geopolymer molecules we write poly(sialate)/polysialate or poly(-sialate-siloxo), a terminology never used in biochemistry. We shall therefore keep our terminology, use it and promote it without any restriction.

Geopolymers comprise following molecular units (or chemical groups):

-Si-O-Si-O- siloxo, poly(siloxo)

-Si-O-Al-O- sialate, poly(sialate)

-Si-O-Al-O-Si-O- sialate-siloxo, poly(sialate-siloxo)

-Si-O-Al-O-Si-O-Si-O- sialate-disiloxo, poly(sialate-disiloxo)

-P-O-P-O- phosphate, poly(phosphate)

-P-O-Si-O-P-O- phospho-siloxo, poly(phospho-siloxo)

-P-O-Si-O-Al-O-P-O- phospho-sialate, poly(phospho-sialate)

-(R)-Si-O-Si-O-(R) organo-siloxo, poly-silicone

-Al-O-P-O- alumino-phospho, poly(alumino-phospho)

-Fe-O-Si-O-Al-O-Si-O- ferro-sialate, poly(ferro-sialate)

Geopolymers are presently developed and applied in 10 main classes of materials:

- Waterglass-based geopolymer, poly(siloxonate), soluble silicate, Si:Al=1:0

- Kaolinite/Hydrosodalite-based geopolymer, poly(sialate) Si:Al=1:1

- Metakaolin MK-750-based geopolymer, poly(sialate-siloxo) Si:Al=2:1

- Calcium-based geopolymer, (Ca, K, Na)-sialate, Si:Al=1, 2, 3

- Rock-based geopolymer, poly(sialate-multisiloxo) 1< Si:Al<5

- Silica-based geopolymer, sialate link and siloxo link in poly(siloxonate) Si:Al>5

- Fly ash-based geopolymer

- Ferro-sialate-based geopolymer

- Phosphate-based geopolymer, AlPO4-based geopolymer

- Organic-mineral geopolymer

## Commercial Applications

There exist a wide variety of potential and existing applications. Some of the geopolymer applications are still in development whereas others are already industrialized and commercialized. They are listed in three major categories:

### 1. Geopolymer Resins and Binders

- Fire-resistant materials, thermal insulation, foams;

- Low-energy ceramic tiles, refractory items, thermal shock refractories;

- High-tech resin systems, paints, binders and grouts;

- Bio-technologies (materials for medicinal applications);

- Foundry industry (resins), tooling for the manufacture of organic fiber composites;

- Composites for infrastructures repair and strengthening, fire-resistant and heat-resistant high-tech carbon-fiber composites for aircraft interior and automobile;

- Radioactive and toxic waste containment;

### 2. Geopolymer Cements and Concretes

- Low-tech building materials (clay bricks),

- Low-$CO_2$ cements and concretes;

### 3. Arts and Archaeology

- Decorative stone artifacts, arts and decoration;

- Cultural heritage, archaeology and history of sciences.

## Geopolymer Resins and Binders

The class of geopolymer materials is described by Davidovits to comprise:

- Metakaolin MK-750-based geopolymer binder:

  chemical formula (Na,K)-(Si-O-Al-O-Si-O-), ratio Si:Al=2 (range 1.5 to 2.5)

- Silica-based geopolymer binder:

  chemical formula (Na,K)-n(Si-O-)-(Si-O-Al-), ratio Si:Al>20 (range 15 to 40).

- Sol-gel-based geopolymer binder (synthetic MK-750):

  chemical formula (Na,K)-(Si-O-Al-O-Si-O-), ratio Si:Al=2

The first geopolymer resin was described in a French patent application filed by J. Davidovits in 1979. The American patent, US 4,349,386, was granted on Sept. 14, 1982 with the title *Mineral Polymers and methods of making them.* It essentially involved the geopolymerization of alkaline soluble silicate [waterglass or (Na,K)-polysiloxonate] with calcined kaolinitic clay (later coined metakaolin MK-750 to highlight the importance of the temperature of calcination, namely 750 °C in this case). In 1985, Kenneth MacKenzie and his team from New-Zealand, discovered the Al(V) coordination of calcined kaolinite (MK-750), describing a "chemical shift intermediate between tetrahedral and octahedral." This had a great input towards a better understanding of its geopolymeric reactivity.

Since 1979, a variety of resins, binders and grouts were developed by the chemical industry, worldwide.

## Potential Utilization for geopolymer Composites Materials

Metakaolin MK-750-based and silica-based geopolymer resins are used to impregnate fibers and fabrics to obtain geopolymer matrix-based fiber composites. These products are fire-resistant; they release no smoke and no toxic fumes. They were tested and recommended by major international institutions such as the American Federal Aviation Administration FAA. FAA selected the carbon-geopolymer composite as the best candidate for the fire-resistant cabin program (1994-1997).

## Fire-resistant Material

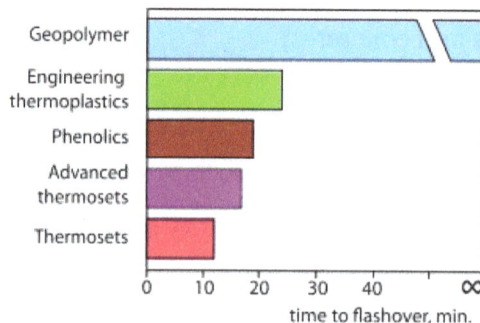

Time to flashover: comparison between organic-matrix and geopolymer-matrix composites

Flashover is a phenomenon unique to compartment fires where incomplete combustion products accumulate at the ceiling and ignite causing total involvement of the compartment materials and signaling the end to human survivability. Consequently, in a compartment fire the time to flashover is the time available for escape and this is the single most important factor in determining the fire hazard of a material or set of materials in a compartment fire. The Federal Aviation Administration has used the time-to-flashover of materials in aircraft cabin tests as the basis for a heat release and heat release rate acceptance criteria for cabin materials for commercial aircraft. The figure shows how the best organic-matrix made of engineering thermoplastics reaches flashover after the 20 minute ignition period and generates appreciable smoke, while the geopolymer-matrix composite will never ignite, reach flashover, or generate any smoke in a compartment fire.

Carbon-geopolymer composite is applied on racing cars around exhaust parts. This technology could be transferred and applied for the mass production of regular automobile parts (corrosion-resistant exhaust pipes and the like) as well as heat shields. A well-known motorcar manufacturer already developed a geopolymer-composite exhaust pipe system.

## Geopolymer Cements

Production of geopolymer cement requires an aluminosilicate precursor material such as metakaolin or fly ash, a user-friendly alkaline reagent (for example, sodium or potassium soluble silicates with a molar ratio MR $SiO_2:M_2O \geq 1.65$, M being Na or K) and water. Room temperature hardening is more readily achieved with the addition of a source of calcium cations, often blast furnace slag.

## Portland Cement Chemistry vs Geopolymer Chemistry

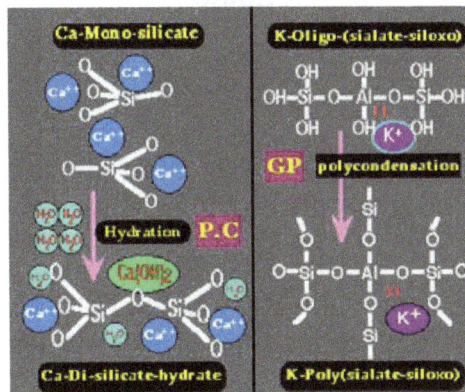

Portland cement chemistry compared to geopolymerization GP

*Left:* hardening of Portland cement (P.C.) through hydration of calcium silicate into calcium silicate hydrate (C-S-H) and portlandite, $Ca(OH)_2$.

*Right:* hardening (setting) of geopolymer cement (GP) through poly-condensation of potassium oligo-(sialate-siloxo) into potassium poly(sialate-siloxo) cross linked network.

## Geopolymer Cement Categories

The categories comprise:

- Slag-based geopolymer cement.

- Rock-based geopolymer cement.

- Fly ash-based geopolymer cement

  o Type 1: alkali-activated fly ash geopolymer.

  o Type 2: slag/fly ash-based geopolymer cement.

- Ferro-sialate-based geopolymer cement.

## Slag-based Geopolymer Cement

*Components*: metakaolin (MK-750) + blast furnace slag + alkali silicate (user-friendly).

*Geopolymeric make-up:* Si:Al = 2 in fact solid solution of Si:Al=1, Ca-poly(di-sialate) (anorthite type) + Si:Al =3 , K-poly(sialate-disiloxo) (orthoclase type) and C-S-H Ca-silicate hydrate.

The first geopolymer cement developed in the 1980s was of the type (K,Na,Ca)-poly(-sialate) (or slag-based geopolymer cement) and resulted from the research developments carried out by Joseph Davidovits and J.L. Sawyer at Lone Star Industries, USA and yielded the invention of Pyrament® cement. The American patent application was filed in 1984 and the patent US 4,509,985 was granted on April 9, 1985 with the title 'Early high-strength mineral polymer'.

## Rock-based Geopolymer Cement

The replacement of a certain amount of MK-750 with selected volcanic tuffs yields geopolymer cement with better properties and less $CO_2$ emission than the simple slag-based geopolymer cement.

*Manufacture components:* metakaolin MK-750, blast furnace slag, volcanic tuffs (calcined or not calcined), mine tailings and alkali silicate (user-friendly).

*Geopolymeric make-up:* Si:Al = 3, in fact solid solution of Si:Al=1 Ca-poly(di-sialate) (anorthite type) + Si:Al =3-5 (Na,K)-poly(silate-multisiloxo) and C-S-H Ca-silicate hydrate.

## Fly Ash-based Geopolymer Cements

Later on, in 1997, building on the works conducted on slag-based geopolymeric cements, on the one hand and on the synthesis of zeolites from fly ashes on the other hand, Silverstrim et al. and van Jaarsveld and van Deventer developed geopolymeric fly ash-based cements. Silverstrim et al. US Patent 5,601,643 was titled 'Fly ash cementitious material and method of making a product'.

## $CO_2$ Emissions During Manufacture

According to the Australian concrete expert B. V. Rangan, the growing worldwide demand for concrete is a great opportunity for the development of geopolymer cements of all types, with their much lower tally of carbon dioxide $CO_2$.

## Coordination Polymer

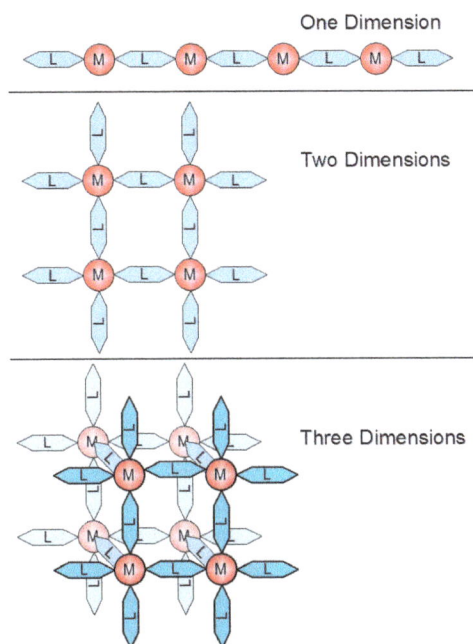

Figure: An illustration of 1- 2- and 3-dimensionality.

A coordination polymer is an inorganic or organometallic polymer structure containing metal cation centers linked by ligands. More formally a coordination polymer is a coordination compound with repeating coordination entities extending in 1, 2, or 3 dimensions.

It can also be described as a polymer whose repeat units are coordination complexes. Coordination polymers contain the subclass coordination networks that are coordination compounds extending, through repeating coordination entities, in 1 dimension, but with cross-links between two or more individual chains, loops, or spiro-links, or a coordination compound extending through repeating coordination entities in 2 or 3 dimensions. A subclass of these are the metal-organic frameworks, or MOFs, that are coordination networks with organic ligands containing potential voids.

Coordination polymers are relevant to many fields such as organic and inorganic chemistry, biochemistry, materials science, electrochemistry, and pharmacology, having many potential applications. This interdisciplinary nature has led to extensive study in the past few decades.

Coordination polymers can be classified in a number of different ways according to their structure and composition. One important classification is referred to as dimensionality. A structure can be determined to be one-, two- or three-dimensional, depending on the number of directions in space the array extends in. A one-dimensional structure extends in a straight line (along the x axis); a two-dimensional structure extends in a plane (two directions, x and y axes); and a three-dimensional structure extends in all three directions (x, y, and z axes). This is depicted in Figure.

## Synthesis and Propagation

Coordination polymers are often prepared by self-assembly, involving crystallization of a metal salt with a ligand. The mechanisms of crystal engineering and molecular self-assembly are relevant.

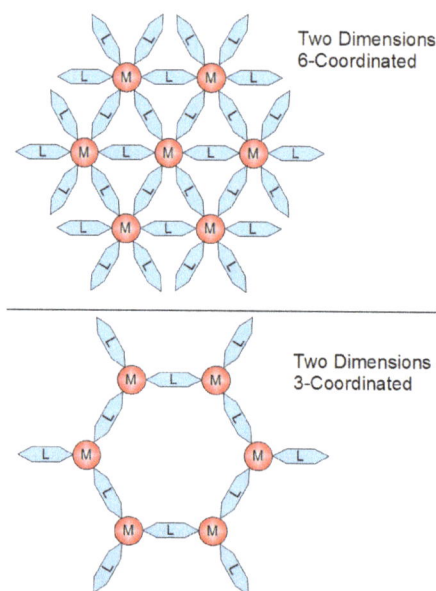

Figure: Shows planar geometries with 3 coordination and 6 coordination.

The synthesis methods utilized to produce coordination polymers are generally the same methods used to grow any crystal. These generally include solvent layering (slow diffusion), slow evaporation, and slow cooling. (Because the main method of characterization of coordination polymers is X-ray crystallography, growing a crystal of sufficient size and quality is important.)

## Intermolecular Forces and Bonding

Forces that determine metal-ligand complexes include van der Waals forces, pi-pi interactions, hydrogen bonding, and stabilization of pi bonds by polarized bonds in addition to the coordination bond formed between the metal and the ligand. These intermolecular forces tend to be weak, with a long equilibrium distance (bond length) compared to covalent bonds. The pi-pi interactions between benzene rings, for example, have energy roughly 5–10 kJ/mol and optimum spacing 3.4–3.8 Ångstroms between parallel faces of the rings.

## Coordination

The crystal structure and dimensionality of the coordination polymer is determined by the functionality of the linker and the coordination geometry of the metal center. Dimensionality is generally driven by the metal center which can have the ability to bond to as many as 16 functional sites on linkers; however this is not always the case as dimensionality can be driven by the linker when the linker bonds to more metal centres than the metal centre does linkers. The highest known coordination number of a coordination polymer is 14, though coordination numbers are most often between 2 and 10. Examples of various coordination numbers are shown in planar geometry in Figure. In Figure the 1D structure is 2-coordinated, the planar is 4-coordinated, and the 3D is 6-coordinated.

## Metal Centers

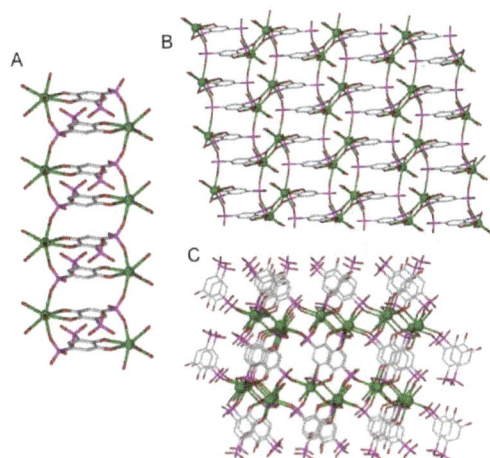

Figure: Three coordination polymers of different dimensionality. All three were made using the same ligand (4,5-dihydroxybenzene-1,3-disulfonate (L)), but different metal cations. All of the metals come from Group 2 on the periodic table (alkaline earth metals) and in this case, dimensionality increases with cation size and polarizability. A. $[Ca(L)(H_2O)_4] \cdot H_2O$ B. $[Sr(L)(H_2O)_4] \cdot H_2O$ C. $[Ba(L)(H_2O)] \cdot H_2O$ In each case, the metal is represented in green.

Metal Centers, often called nodes or hubs, bond to a specific number of linkers at well defined angles. The number of linkers bound to a node is known as the coordination number, which, along with the angles they are held at, determines the dimensionality of the structure. The coordination number and coordination geometry of a metal center is determined by the nonuniform distribution of electron density around it, and in general the coordination number increases with cation size. Several models, most notably hybridization model and molecular orbital theory, use the Schrödinger equation to predict and explain coordination geometry, however this is difficult in part because of the complex effect of environment on electron density distribution.

## Transition Metals

Transition metals are commonly used as nodes. Partially filled d orbitals, either in the atom or ion, can hybridize differently depending on environment. This electronic structure causes some of them to exhibit multiple coordination geometries, particularly copper and gold ions which as neutral atoms have full d-orbitals in their outer shells.

## Lanthanides

Lanthanides are large atoms with coordination numbers varying from 7 to 14. Their coordination environment can be difficult to predict, making them challenging to use as nodes. They offer the possibility of incorporating luminescent components.

## Alkali Metals and Alkaline Earth Metals

Alkali metals and alkaline earth metals exist as stable cations. Alkali metals readily form cations with stable valence shells, giving them different coordination behavior than lanthanides and transition metals. They are strongly effected by the counterion from the salt used in synthesis, which is difficult to avoid. The coordination polymers shown in figure are all group two metals. In this case, the dimensionality of these structures increases as the radius of the metal increases down the group (from calcium to strontium to barium).

## Ligands

In most coordination polymers, a ligand (atom or group of atoms) will formally donate a lone pair of electrons to a metal cation and form a coordination complex via a Lewis acid/ base relationship (Lewis acids and bases). Coordination polymers are formed when a ligand has the ability to form multiple coordination bonds and act as a bridge between multiple metal centers. Ligands that can form one coordination bond are referred to as monodentate, but those which form multiple coordination bonds, which could lead to coordination polymers are called polydentate. Polydentate ligands are particularly important because it is through ligands that connect multiple metal centers together that an infinite array is formed. Polydentate ligands can also form multiple bonds to the same metal (which is called chelation). Monodentate ligands are also referred to as terminal because they do not offer a place for the network to continue. Often, coordination polymers will consist of a combination of poly- and monodentate, bridging, chelating, and terminal ligands.

## Chemical Composition

Almost any type of atom with a lone pair of electrons can be incorporated into a ligand. Ligands that are commonly found in coordination polymers include polypyridines, phenanthrolines, hydroxyquinolines and polycarboxylates. Oxygen and nitrogen atoms are commonly encountered as binding sites, but other atoms, such as sulfur and phosphorus, have been observed.

Ligands and metal cations tend to follow hard soft acid base theory (HSAB) trends. This means that larger, more polarizable soft metals will coordinate more readily with larger more polarizable soft ligands, and small, non-polarizable, hard metals coordinate to small, non-polarizable, hard ligands.

## Structural Orientation

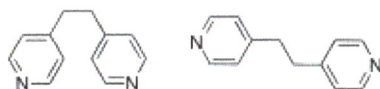

Gauche          Anti

1,2-Bis(4-pyridyl)ethane is a flexible ligand, which can exist in either gauche or anti conformations.

Ligands can be flexible or rigid. A rigid ligand is one that has no freedom to rotate around bonds or reorient within a structure. Flexible ligands can bend, rotate around bonds, and reorient themselves. These different conformations create more variety in the structure. There are examples of coordination polymers that include two configurations of the same ligand within one structure, as well as two separate structures where the only difference between them is ligand orientation.

## Ligand Length

A length of the ligand can be an important factor in determining possibility for formation of a polymeric structure versus non-polymeric (mono- or oligomeric) structures.

## Other Factors

## Counterion

Besides metal and ligand choice, there are many other factors that affect the structure of the coordination polymer. For example, most metal centers are positively charged ions which exist as salts. The counterion in the salt can affect the overall structure. For example, silver salts such as $AgNO_3$, $AgBF_4$, $AgClO_4$, $AgPF_6$, $AgAsF_6$ and $AgSbF_6$ are all crystallized with the same ligand, the structures vary in terms of the coordination environment of the metal, as well as the dimensionality of the entire coordination polymer.

## Crystallization Environment

Additionally, variations in the crystallization environment can also change the structure. Changes in pH, exposure to light, or changes in temperature can all change the resulting structure. Influences on the structure based on changes in crystallization environment are determined on a case by case basis.

## Guest Molecules

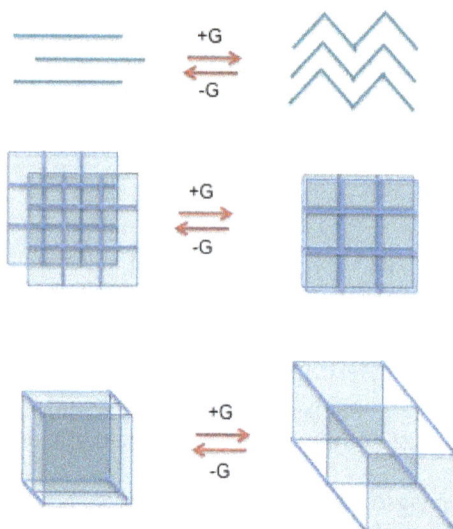

The addition and removal of guest molecules can have a large effect on the resulting structure of a coordination polymer. A few examples are (top) change of a linear 1D chain to a zigzag pattern, (middle) staggered 2D sheets to stacked, and (bottom) 3D cubes become more widely spaced.

The structure of coordination polymers often incorporates empty space in the form of pores or channels. This empty space is thermodynamically unfavorable. In order to stabilize the structure and prevent collapse, the pores or channels are often occupied by guest molecules. Guest molecules do not form bonds with the surrounding lattice, but sometimes interact via intermolecular forces, such as hydrogen bonding or pi stacking. Most often, the guest molecule will be the solvent that the coordination polymer was crystallized in, but can really be anything (other salts present, atmospheric gases such as oxygen, nitrogen, carbon dioxide, etc.) The presence of the guest molecule can sometimes influence the structure by supporting a pore or channel, where otherwise none would exist.

## Applications

Coordination polymers are commercialized as dyes. Particularly useful are derivatives of aminophenol. Metal complex dyes using copper or chromium are commonly used for producing dull colors. Tridentate ligand dyes are useful because they are more stable than their bi- or mono-dentate counterparts.

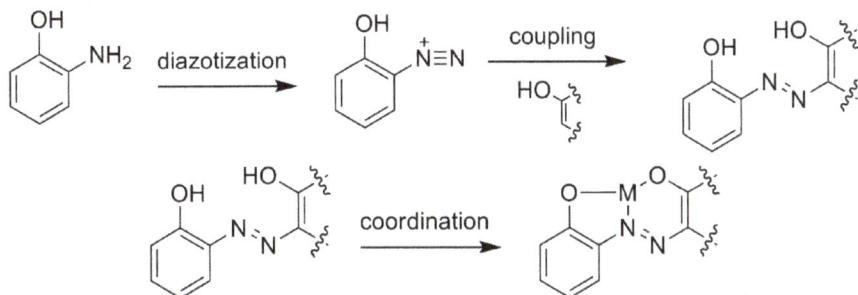

One of the early commercialized coordination polymers are the Hofmann compounds, which have the formula $Ni(CN)_4Ni(NH_3)_2$. These materials crystallize with small aromatic guests (benzene, certain xylenes), and this selectivity has been exploited commercially for the separation of these hydrocarbons.

## Conductive Polymers

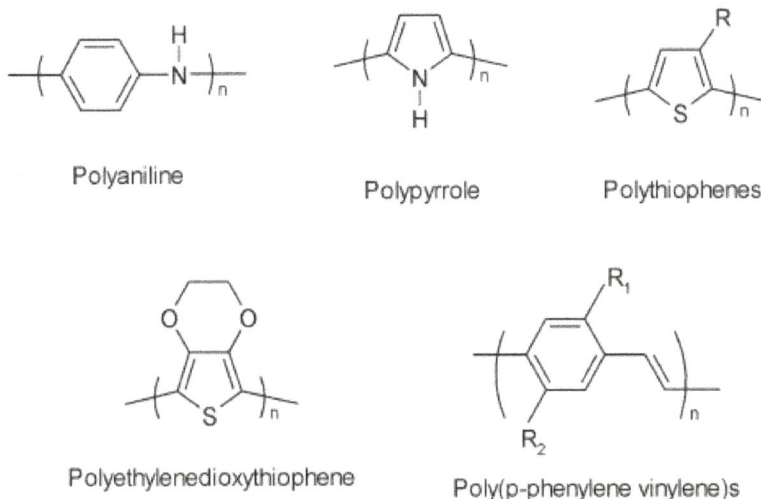

Polyaniline

Polypyrrole

Polythiophenes

Polyethylenedioxythiophene

Poly(p-phenylene vinylene)s

Conductive polymers are organic polymers that conduct electricity. It was first described in Polyaniline in mid-19th century by Henry Lethe by, and developed gradually since then.

Conducting polymers (ICPs) exist that have alternating single and double bonds along the polymer backbone (conjugated bonds) or that are composed of aromatic rings such as phenylene, naphthalene, anthracene, pyrrole, and thiophene which are connected to one another through carbon-carbon single bonds.

The first polymer with significant conductivity synthesized was polyacetylene (polyethyne). Its electrical conductivity was discovered by Hideki Shirakawa, Alan Heeger, and Alan MacDiarmid who received the Nobel Prize in Chemistry in 2000 for this discovery. They synthesized this polymer for the first time in the year 1974 when they prepared polyacetylene as a silvery film from acetylene, using a Ziegler-Natta catalyst. Despite its metallic appearance, the first attempt did not yield a very conductive polymer. However, three years later, they discovered that oxidation with halogen vapor produces a much more conductive polyacetylene film.[1] Its conductivity was significantly higher than any other previously known conductive polymer. This discovery started the development of many other conductive organic polymers.

The conductivity of non-doped, conjugated polymers such as polyacetylene is due to the existence of a conducting band similar to a metal. In a conjugated polymer three of the four valence electrons from strong σ bonds through sp² hybridization where elctrons are strongly localized. The remaining unpaired electron of each carbon atom remains in a $p_z$ orbital. It overlaps with a neighboring $p_z$ orbital to form a π bond. The π electrons of these conjugated $p_z$ orbitals overlap to form an extended $p_z$ orbital system through which electrons can move freely (delocalization of π electrons). However, non-doped polymers have a rather low conductivity. Only when an electron is removed from the valence band by oxidation (p-doping) or is added to the conducting band by reduction (n-doping) does the polymer become highly conductive. The four main methods of doping are:

- Redox p-doping: Some of the π-bonds are oxidized by treating the polymer with an oxidizing agent such as iodine, chlorine, arsenic pentafluoride etc.

- Redox n-doping[2]: Some of the π-bonds are reduced by treating the polymer with a reducing agents such as lithium, and sodium naphthaline.

- Electrochemical p- and n-doping: Doping is achieved by cathodic reduction (p) or by anodic oxidation (n).

- Photo-Induced Doping: The polymer is exposed to high energy radiation that allows electrons to jump to the conducting band. In this case, the positive and negative charges are localized over a few bonds.

Doping increases the conductivity by many orders of magnitude. Values as high as $10^2$ - $10^4$ S/m have been reported. Another method to increase conductivity is mechanical alignment of the polymer chains. In the case of polyacetylene, conductivities as high as $10^5$ S/m have been found which is still several magnitudes lower than the conductivity of silver and copper ($10^8$ S/m) but more than sufficient for electronic applications such as polymer-based transistors, light-emitting diodes and lasers.

The table below lists typical conductivities of some common conjugated polymers and their repeat units. The actual conductivity not only depends on the structure and morphology of the polymer but also on the type of dopant and its concentration.

| ELECTRICAL CONDUCTIVITY OF SOME CONDUCTIVE POLYMERS | | |
|---|---|---|
| Compound | Repeating Unit | Conductivity (S cm-1) |
| trans-Polyacetylene | | 103 - 105 |
| Polythiophene | | 103 |
| Polypyrrole | | 102 - 7.5 · 103 |
| Poly(p-phenylene) | | 102 - 103 |
| Polyaniline | | 2 · 102 |
| Poly(p-phenylene vinylene) | | 2 · 104 |

## Polyphosphazene

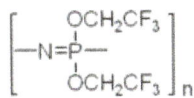

$$\left[ -N=P- \begin{array}{c} OCH_2CF_3 \\ | \\ | \\ OCH_2CF_3 \end{array} \right]_n$$

Hydrophobic film-, fiber-, and membrane-forming material

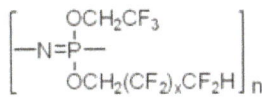

$$\left[ -N=P- \begin{array}{c} OCH_2CF_3 \\ | \\ | \\ OCH_2(CF_2)_xCF_2H \end{array} \right]_n$$

Hydrophobic elastomer

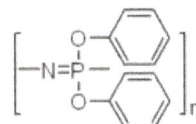

$$\left[ -N=P- \begin{array}{c} O- \\ | \\ | \\ O- \end{array} \right]_n$$

Hydrophobic film- and fiber-forming material

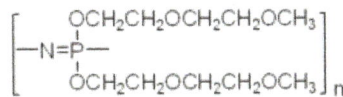

$$\left[ -N=P- \begin{array}{c} OCH_2CH_2OCH_2CH_2OCH_3 \\ | \\ | \\ OCH_2CH_2OCH_2CH_2OCH_3 \end{array} \right]_n$$

Water soluble polymer and solid polymer electrolyte

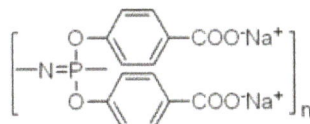

$$\left[ -N=P- \begin{array}{c} O- \bigcirc -COO^-Na^+ \\ | \\ | \\ O- \bigcirc -COO^-Na^+ \end{array} \right]_n$$

Water soluble polymer

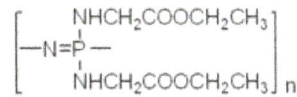

$$\left[ -N=P- \begin{array}{c} NHCH_2COOCH_2CH_3 \\ | \\ | \\ NHCH_2COOCH_2CH_3 \end{array} \right]_n$$

Bioerodible polymer

Polyphosphazenes comprise the largest class of inorganic–organic polymers with more than 700 variations which have very interesting properties and commercially promising applications. Synthesis of the first polyphosphazene has been reported in 1964.

The backbone of these polymers consists of alternating phosphorus and nitrogen atoms which each one attaches to organic or inorganic side groups. The most interesting feature of these compounds is their synthesis procedure which allows changing of the side groups to obtain a wide variety of products with a broad range of properties. It is consisted of elastomers to glasses, hydrophilic to hydrophobic, bioinert to bioactive materials, and electrical conductors to insulators. Thus, considering the structure-property correlations, one can even predict the properties of the compound yet to be prepared. The properties-structure relationship for some typical polyphosphazenes is shown in table.

As crystallization is a consequence of molecular symmetry, the microcrystallinity is only observed in polyphosphazene with a single type of substituent group. When different types of substitute groups are present, the required conditions with respect to symmetry do not meet. The aminophosphazene polymers are amorphous probably due to hydrogen bonding, for example. The elastomeric properties are exhibited when small or flexible substituents, e.g., linear side groups such as $-OC_2H_5$, $-OCH_3-$ are present. Therefore, these polymers are used as low temperature elastomers.

Polyphosphazenes with trifluoroethoxy, alkyl, aryl, and organosilicon are also used for this application. Also, polyorganophosphazenes are resistant to hydrolysis.

As, the molecular-property relationships play an important role in producing a wide range of advanced materials, some general relationships can be summarized as follows:

- Crystalline versus amorphous polymers: As mentioned earlier, presence of one type of side group results in molecular symmetry, microcrystallinity, and mixed substitutions end up with amorphous structure.

- Hydrophobic versus hydrophilic polymers: The presence of a hydrophilic side group such as $-NHCH_3$, results in solubility of the polymer in water and side group such as $-OCH_2CF_3$ which is hydrophobic in nature leads to water repellency.

- Water stability versus water erodability: Most of the poly(organophosphazenes) are stable in water but, polymers containing amino acid ester side groups are unstable in water. Also, water-soluble polyphosphazenes with aminoalcoholic functions have been prepared by reaction of aminoalcohol with polydichlorophosphazene. Poly[di(carboxylatophenoxy)phosphazene is a water soluble polymer and is used in aqueous microencapsulation.

- High Tg versus low Tg polymers: For different side groups as shown in table, Tg can move up or down.

Far about 300 grades of polyphosphazenes have been made, and most of them are functionalized polyphosphazenes, which are bonded to other mole- cules or to other polymers to make novel materials. Polyphosphazenes form new promising engineering materials with a large number of possible different structures for specific applications. Some of these polymers are very hard, stiff, and have very good.

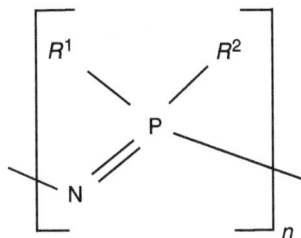

The chemical structure of backbone of polyphos- phazene

Table: Relationship of properties-structure for some polyphosphazenes

| Formula | Properties | Tg=$^{\circ}$C | Tm=$^{\circ}$C |
|---|---|---|---|
| $[NP(NHC_6H_5)_2]_n$ | Glass | +91 | – |
| $[NP(OC_6H_5)(OC_6H_4C_6H_5\text{-}p)]_n$ | Glass | +43 | – |
| $[NP(OC_6H_5)(OC_6H_4C_6H_5\text{-}o)]_n$ | Glass | +24 | – |
| $[NP(OC_6H_4COOEt)_2]_n$ | Microcrystalline thermoplastic (films) | +7.5 | +127 |
| $[NP(OC_6H_5)_2]_n$ | Microcrystalline thermoplastic (films, fibers) | -8 | +390 |
| $[NP(OC_6H_5)(OC_6H_4C_2H_5)]_n$ | Elastomer | -10 | – |
| $[NP(O(CH_2)_8CH_3)_2]_n$ | Elastomer | -56 | – |
| $[NP(OCH_2CF_3)_2]n$ | Microcrystalline thermoplastic (films, fibers) | -66 | +242 |
| $[NP(OCH_3)_2]_n$ | Elastomer | -76 | – |
| $[NP(OC_2H_5)_2]_n$ | Elastomer | -84 | – |
| $[NP(OC_3H_7)_2]_n$ | Elastomer | -100 | – |
| $[NP(OCH_2CH_2CH_2CH_3)_2]n$ | Elastomer | -105 | – |

resistance to organic chemicals. Furthermore, this class of materials combines high temperature stability and low temperature flexibility with good tough- ness and inherent flame retardance.

These polymers are more resistant to thermo-oxida- tive decomposition than organic polymers. Several polyphosphazene materials have thermal stability above 400oC and are resistant to degradation in acidic media. DSC technique has been used to study thermal behavior of polyphosphazene homopolymers. The results show that polyphosphazenes have a single glass transition temperature and two first order phase transitions with a large mesophase temperature. When these polymers are obtained in solution, they exhibit 3D crystallinity and by heating above their mesophase transition they show 2D structure and when they are cooled to room temperature, their crystallinity will be at very high level. In these polyphosphazenes, the glass transition temperature is almost independent of the level of crystallinity. The nature and magnitude of mesophase transition temperature and melt tempera- ture depend upon processing conditions. Polypho- sphazenes, especially those which contain aryloxy side groups, have very good temperature stability.

## Polyphosphazene Applications

Although polyphosphazenes can be customized for virtually any application, most activity can be grouped into four segments: Fuel Cell , Medical, High Performance, and Membrane.

Fuel Cell: A fuel cell is a device that produces electricity by efficient electrochemical conversion of fuel. Polyphosphazene is currently the highest performing membrane material for Methanol based Proton-Exchange Membrane (PEM) Fuel Cells. This fuel cell type is ideal for miniature power supply and is a leading candidate for automotive applications. Specially-fabricated polyphosphazene membranes possess high proton conductivity, slow methanol transport, and stability (thermal, mechanical, and chemical). They out-perform most other membrane materials in methanol environments and have demonstrated minimal methanol crossover, better thermal performance in high-temperature environments, and minimal chemical and mechanical degradation.

Medical: Polyphosphazenes make ideal medical polymers because of their biocompatibility, capability for intricate customization, high affinity for water, and the ability to accept grafts of influencing substituents. Medical applications include: Drug Delivery, Biological Membranes, Coatings, and polymeric medical devices and components such as prosthetics and implants. Drug delivery systems utilize polyphosphazene's ability to accept grafts of active components to create a 'carrier-molecule' for the pharmaceutical. In addition, a substituent capable of influencing the behavior of the chain (i.e.: accelerating hydrolysis), can be added, enabling the rate of release to be a function of the polymer characteristics and substituent ratios. Phosphazenes are also used as surface coatings and composite materials. Derivatives have been synthesized to be inert in biological media and are known to function well subcutaneously.

High Performance: Polyphosphazenes are used as flame retardants, additives, performance polymers, and in specialty applications. The exceptional performance of polyphosphazene derivatives under extreme temperature conditions, their inertness to chemical environments, and their non-flammability, make them premiere materials for applications in hostile landscapes. Polyphosphazene derivatives have been synthesized that exhibit a low Tg (-65oC) with a thermal operating temperature range from -90 to 200oC. Their excellent combustion behavior such as low smoke emission, no corrosiveness, low toxicity of gases, and their ability to withstand a diversity of hostile chemical environments make them premiere materials for uses in fluoroelastomeric seals, gaskets, O-rings, and in insulating foams. Other products include specialty rubbers, flame resistant materials, polymer conductors, lubricants, liquid crystal polymers, catalysis, paints, adhesives, photocuring polymers, self-stabilized polymers, and additives.

Membrane: Polyphosphazenes are being used to make membranes more thermally, mechanically, and chemically stable, as well as to enhance selectivity and overall performance. They are mainly used in electrodialysis, microfiltration, ultrafiltration, and reverse osmosis applications. They can be engineered as membranes per se or in hollow tube and depth media. Phosphazene membranes have shown immense potential in water purification, fuel-gas technologies, air enrichment, sour gas purification, and in the separation of mixtures of alcohols and various organic compounds and ions.

# Ionomer

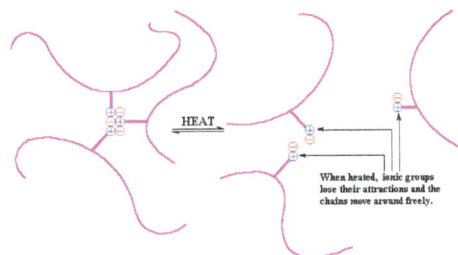

HEAT

When heated, ionic groups lose their attractions and the chains move around freely.

Ionomers are synthetic polyelectrolytes that consist of both electrically neutral and ionized groups that are randomly and or regularly distributed along the polymer backbone. They can be divided into *polycations*, *polyanions* and *polyampholytes*. Like ordinary polyelectrolytes, they can bear one or more charges depending on the pH value. However, per definition, their ion content is usually no more than 10 to 15% and, thus, most ionomers are insolube or only slightly soluble in water. One important characteristic of ionomers is the strong molecular aggregation of the ion-carrying groups to ion rich domains or ion clusters which act as physical crosslinks. When heated, the ionic bonds and clusters dissolve and when cooled, they reform. This gives ionomers a unique structure and behavior. At low temperatures they behave like crosslinked polymers (elastomers) and at elevated temperatures like ordinary thermoplastics.

The majority of ionomers studied have a polyvinyl or polydiene backbone and carry anionic groups (ionomers) with $Na^+$, and $Zn^{2+}$ as counter ions. The three most important ionized groups are carboxylate ($-COO^-$), sulfonate ($-SO_3^-$) and phosphonate ($-PO_3^{2-}$) which differ in the strength of the ionic interaction. Usually, less than 80% of the acid groups are neutralized by a base. The remaining acid groups provide sites for hydrogen bonding between neighboring moelcules which are weaker than ionic bonds but stronger than secondary bonds. Thus, increasing the acid and ion content in the polymer results higher mechanical strength, modulus, and toughness. Carboxylate ionomers with an olefin backbone are by far the most common ionic polymers followed by sulfonate ionomers with a styrene backbone. Both types of ionomers are produced on a commercial scale and find many applications. Other backbones that have been studied include polybutadiene, polyacrylate, polymethacrylate, polyisoprene, and polytetrafluorethylene.

Ionomers can be either synthesized by copolymerization of ionic and neutral monomers or by chemical modification of electrically neutral polymers. The majority of ionomers are produced via free-radical copolymerization. For example, ethylene is copolymerized with methacrylic and/or acrylic acid by a high pressure process similar to low-density polyethylene. These ionic copolymers have typically a low melting point, improved toughness/flexibility and mechanical strength and when used as films, posess a much higher clarity and gloss and provide superior hot tack, seal strength and puncture resistance than unmodified polyethylene film. Another important class of carboxylate ionomers are copolymers of acrylates with acrylic and/or methacrylic acid, produced by free-radical emulsion or solution copolymerization. These ionomers are used as pressure-sensitive adhesives. A commercially important sulfonate ionomer is sulfonated tetrafluoroethylene (Nafion) which is produced by free-radical copolymerization of tetrafluoroethylene (TFE is the monomer of Teflon) with a perfluorinated vinyl ether sulfonyl fluoride co-monomer. The copolymers carries perfluoroether pendant side chains terminated by sulfonic acid groups. These ionomers have excellent chemical and thermal stability and can absorb large amounts of water. They are often used as ion-selective membranes. An ionomer that is produced via chemical modification is (lightly) sulfonated polystyrene. These ionomers are obtained by treating polystyrene with sulfonating agents such as acetyl sulfate in chlorinated solvents (post-sulfonation). This method can also be employed to produce block-copolymers with lightly sulfonated styrene blocks. The sulfonated ionomers form physical crosslinks that approach the strength of covalent links which is desirable in many elastomeric applications.

## Commercial Ionomers

The largest volume ionomers are copolymers of ethylene and acrylic and/or methacrylic acid. Commercial grades of ethylene (meth)acrylic acid (EAA, EMAA) are available from DuPont (Surlyn®, Nucrel®), SK Global Chemical[1] (Primacor™), and Ineos, (Eltex®).

Perfluorinated sulfonic acid ionomers (PFSA) are sold by DuPont (Nafion), Solfay (Aquivion), Chemours, and 3M.

| Ionomers | Structure of Repeat Unit |
|---|---|
| Poly(ethylene-co-methacrylic acid) Ionomer (Na, Zn) | |
| Poly(ethylene-co-acrylic acid) Ionomer (Na, Zn) | |
| Perfluorinated sulfonic acid ionomers (PFSA), Nafion | |

## Applications

The by far most important ionomer is ethylene acrylic acid copolymer (EAA) which is sold under the tradename Surlyn by DuPont. It is frequently used as a food packaging material and as a tie-layer (compatibilizer) in multi-layer films. Other important applications include coatings and surface films for golf balls, sports equipment, and for overmolded (cosmetic) bottles.

Perfluorinated sulfonic acid ionomers (Nafion) are often used as ion-selective membranes. Important applications include cation exchange membrane for fuel cells, PEM water electroyzers, separators for redox flow batteries, electrodialysis, and electrochemical hydrogen compressors.

## References

- Bernstein, Jeremy; Paul M. Fishbane; Stephen G. Gasiorowicz (April 3, 2000). Modern Physics. Prentice-Hall. p. 624. ISBN 978-0-13-955311-0

- Inorganic-polymer, materialsparts-and-finishes: what-when-how.com, Retrieved 17 July 2018

- Meinhold, R. H.; MacKenzie, K. J. D.; Brown, I. W. M. (1985). "Thermal reactions of kaolinite studied by solid state 27-Al and 29-Si NMR". Journal of Materials Science Letters. 4 (2): 163–166. doi:10.1007/BF00728065. ISSN 0261-8028

- Polyphosphazene: technically.com, Retrieved 27 May 2018

- Batten, Stuart R. (2008). Coordination Polymers: Design, Analysis and Application. RSC Publishing. pp. 297–307, 396–407. doi:10.1039/9781847558862. ISBN 978-0-85404-837-3

- Barbosa, V.F.F; MacKenzie, K.J.D. and Thaumaturgo, C., (2000), Synthesis and characterization of materials based on inorganic polymers of alumina and silica: sodium polysialate polymers, Intern. Journal of Inorganic Materials, 2, pp. 309–317

# Fluoropolymers

A fluorocarbon-based polymer which has multiple carbon-fluorine bonds is termed as a fluoropolymer. It has a high resistance to acids, bases and solvents. Polytetrafluoro-ethylene, ECTFE perfluoroalkoxy alkanes, polyvinylidene fluoride, etc. are some of the important fluoropolymers covered in this chapter.

A fluoropolymer is a plastic material based in fluorocarbon containing carbon fluorine molecules. It is a special type of polymer that possesses special properties. The electro-negative ion fluoride in a fluoropolymer gives it strong carbon-fluoride bonds, which makes it non-sticky.

Fluoropolymers possess highly industrial, demand-specific characteristics, such as:

- Inertness to most chemicals
- Excellent non-stick
- Resistance to high temperatures
- Resistance to galling
- Abrasion resistance
- Waterproof
- Extremely low coefficients of friction
- Excellent dielectric properties, which are relatively insensitive to temperature and power frequency

The combination of these characteristics makes fluoropolymers good for use in the electronics, automotive and aero industries; pipe and chemical processing equipment, and non-stick coatings for cookware and other applications.

Fluoropolymers are made from thermoplastic resins similar to polyethylene, in which some of the hydrogen atoms attached to the carbon chain are replaced by fluorine or fluorinated alkyl groups.

Fluoropolymer coatings protect the products that they are applied to which makes the product last longer and function better through its lifespan. For example:

- Fluoropolymer coatings are applied to the wiring insulation in airplanes. This application protects the wiring and reduces the risk of fire on aircrafts. The next time you are flying across the country, you can rest easy knowing that your plane is protected from fire thanks to fluoropolymer coatings.

- Fluoropolymers are used in many industrial settings. Printing and packaging factories use fluoropolymers to coat their ink trays, rolls, and frames. The coating helps moving parts to not stick to items that pass through, and it also makes them easier to clean.

- Fluoropolymer coatings are also widely used in the food industry. The application of fluoropolymer coatings to molds and trays aids in the production of bread, candy, cheese, and other products by preventing sticking and making the molds and trays easy to clean.

- A big demand for fluoropolymer coatings comes from the automotive industry. Fluoropolymer coatings help to prevent friction and corrosion on car parts. They are particularly useful in components like ball bearings and gears, which must withstand a lot of wear and tear. This application extends the lifespan of your car's parts.

## Polytetrafluoroethylene

Polytetrafluoroethylene (PTFE) is a fluoropolymer, classified among thermoplastics. It was discovered by Dr. Roy J. Plunkett at the DuPont industry and it is commercially named as Teflon. PTFE is polymerized from the monomer tetrafluoroethylene (TFE). It possesses the C-F bond in the PTFE formulation, having the molecular formula [($CF_2$-$CF_2$)n]. PTFE is a high molecular weight compound because of the strongly bonded fluorine atoms and exhibits semi-crystalline nature. The strong C-F bond does not lead PTFE to react with any other compounds. The carbon backbone is tightly bonded to the fluorine atom could be the reason. Hence, interfacing PTFE in the micro to macro components require improvement of its property and performance-related studies to overcome the technical hindrance. Many peculiar properties such as high thermal conductivity in composite form, mechanical strength, hydrophobicity, and chemical inertness have been manifested by PTFE. The average melting point of PTFE is 325 to 335 °C hence PTFE has been classified among thermoplastics because of the high thermal resistance and high operating temperature. Due to its physical and chemical properties, PTFE has been considered as a functional polymeric material in industrial domains. It has been preferred as a nonstick coating material to withstand high-temperature cycles. PTFE is an engineering polymer in terms of mechanical uses such as lubrication, bearing balls, and polymeric gears. The dielectric property of PTFE offers insulating applications such as coatings for cables, electrets for storing electrical charges and making of printed circuit boards. In clinical applications, the PTFE coatings preferred for implants, stents, and biomedical instrumentations due to the inert characteristics. Unlike other polymers, casting of PTFE by common methods is a challenging task but so far different methods were employed to attain the shape and form. PTFE has been graded with additives and filled with micro/nanoparticles to form a composite which provides a convenient pathway for tuning the properties.

## Properties of PTFE

PTFE is ideal for performance due to unique properties. The molecular structure of PTFE is depicted in Fig. The properties of PTFE widely spread in all the branches and being used for a variety of applications. Various properties of PTFE are represented in Fig. Reports on various properties of PTFE have been discussed in this topic.

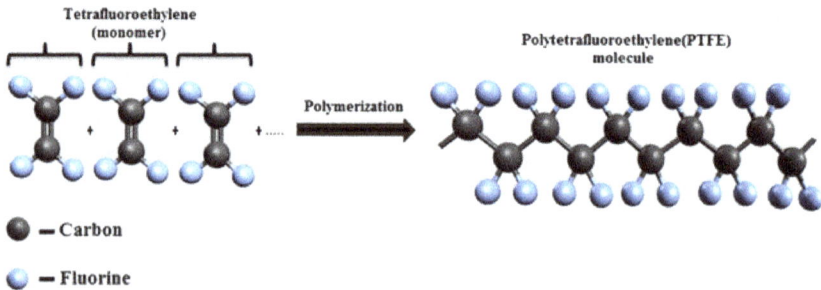

Fig: Molecular structure of polytetrafluoroethylene (PTFE)

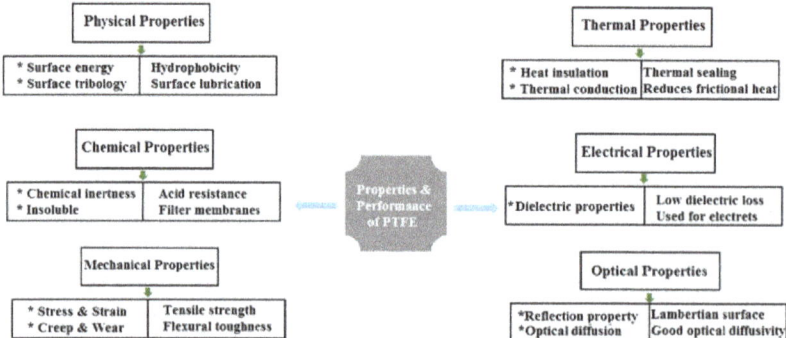

Fig: The various properties and the role of PTFE performance

## Physical Properties of PTFE

## Barrier Properties

PTFE demonstrated superior hydrophobic nature due to the low surface energy. Wrinkled superhydrophobic surfaces, fabricated from two forms of PTFE exhibit the durable and excellent barrier properties as the roll-off angle of the surfaces tend to be very low. The contact angle for single-scale wrinkled PTFE and hierarchical wrinkled PTFE surface was measured at 163° and 172° that has been possessing higher magnitude. Surface modification of PTFE from hydrophobic to hydrophilic property was optimized by the addition of chemical agents amino ($-NH_2$), carboxyl (-COOH) and sulfonic acid ($-SO_3H$). On microfiltration analysis, membranes of PTFE adhered with hydrophilic agent's shows good microfiltration property. The property of plasma modified PTFE is more effective in high-performance direct contact membrane distillation (DCMD). PTFE has been treated with a plasma to obtain pore on the surfaces. Surface morphology study reveals the appearance of parallel pore layers during plasma treatment. Plasma treatment deploys contact angle as a function of treatment time. The bipolar Argon plasma treatment of PTFE also supports the same as with plasma treatment there is an increase in surface free energy. Even though the low surface energy property of virgin PTFE is useful but somehow it is difficult to blend or grind with other polymers. In such a case, the modification of surface is achieved using high-energy irradiation which is in connection with degradation process. Further examination revealed that under irradiation, the PTFE compound has low molecular weight and lower hydrophobicity.

A work on extended PTFE tape demonstrated the increase in water contact angle in a feasible manner. PTFE was stretched using a mechanical device for different ratio of extension. It was portrayed that the increasing extension ratio significantly increases the water contact angle of the surface. The water-repellent property of the surface was mainly due to the decrease in the density of PTFE tape under stretching. More precisely the reason for higher contact angle was the alignment of PTFE microforms on the tape.

The composition of inorganic fullerene-like tungsten disulfide (IF-WS$_2$) nanoparticles and PTFE improved the hydrophobic property. The drastic change in surface roughness of IF-WS$_2$/PTFE was revealed by atomic force microscope (AFM) images. Such key factor was the reason for the increase in hydrophobicity. The different contact angles reported for PTFE and PTFE composites are graphically shown in Figure.

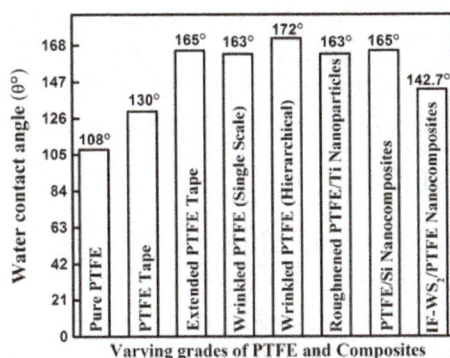

Figure: Contact angle is shown for different grades of
PTFE and PTFE composites and their hydrophobic nature

## Tribological Property of PTFE Surface

- Surface friction of PTFE

  Extensive studies have been done on the friction property of PTFE because of interesting low friction coefficient. Friction occurs due to the relative motion of the surface. A virgin PTFE reveals the ultimate friction resistance property therefore optimized for different types of lubrication. As a function of glass fiber, carbon, and graphite loading, there has been a strong influence over friction properties. A wear mechanism was reported for metal precursor-based PTFE composites. Nano-sized PTFE particles were filled in nickel (Ni) and phosphorous (P) coatings. The work revealed the considerable change in friction coefficient ($\mu$) of Ni-P/PTFE coatings when subjected to wear test. Comparatively, Ni-P/PTFE coating exhibited low wear resistance than Ni-P coating because of the presence of PTFE particles.

- Surface wear of PTFE

Wear is the most important property of PTFE among surface properties. In general, wear is also associated with mechanical properties, the parameters of wear include weight load, velocity, temperature, contact area, and sliding distance. A familiar method known as Pin-on-Disc setup was used to analyze the wear behavior. This test showed that the friction coefficient for virgin PTFE decreased with the increase in loading of carbon and bronze, increased the wear resistance whereas the friction coefficient was affected slightly. PTFE with filler loadings was effectively have good wear resistance when bearing the weight load over the surface.

- Surface lubrication of PTFE

The performance of PTFE as self-lubricant bearings was well known and often examined for its excellent sliding behavior. The mitigation of London dispersive forces in PTFE is due to the highly electronegative fluorine atoms. Furthermore, this property of PTFE was thoroughly examined to improve for high efficiency. In PTFE compound, the fluorine atoms are very close, forming a smooth and cylindrical surface so as the other molecules sliding over easily. In tribological view, PTFE is the topmost material preferred among all.

- Abrasion property of PTFE

Abrasion property of PTFE interlinked with wear rate and friction coefficient. Pure PTFE compound are good in abrasion resistance but fixing it on the surface is the challenging task. Glass fiber (GF) and carbon fiber (CF) filled PTFE were tested for the abrasion resistance capacity. The abrasiveness and surface morphology of the worn surfaces of GF/PTFE and CF/PTFE was studied using scanning electron microscope (SEM). The wear volume was certainly lost in GF/PTFE than CF/PTFE. Under various weight loads, CF/PTFE poses better abrasion resistance because of the adhesion of carbon fibers with the PTFE matrix. Although PTFE possesses lower friction property than any other polymer, the addition of filler makes it suitable for interfacing with good friction resistance.

- Mechanical properties of PTFE

Tensile, hardness, stress, and strain tests on PTFE

The mechanical property of PTFE deals with the study of tensile strength, stress and strain, ductility, hardness, and molding ability. PTFE is ductile in nature and obviously remains low in mechanical phase when compared to other polymers but PTFE has a good advantage in constructing mechanical device parts by loading filler components. Compression test on two grades of PTFE exhibited good mechanistic performance. Significantly the mechanical properties are affected by temperature hence the samples of PTFE were also tested with the load of 50% at a temperature varying from – 198 to 200 °C. During deformation, PTFE undergoes a structural change of approximately 30% in comparison with metals which are less than 10%. The rearrangement of molecules due to strain is temporary because of the viscoelastic nature of polymers and permanent damage when it reaches the physical aging.

Generally, the unfilled PTFE exhibits very poor flexural properties. An improvement over mechanical property has been studied in detail for the composite material Polyamide6 (PA6)/PTFE. Flexural and tensile properties test were conducted for different PA6 content. The samples were analyzed by keeping constant load for five specimens of different magnitudes and the morphology was observed using SEM. Under stress, the deformation of PTFE occurs and improves the flexural toughness due to the absorption of energy. Results showed that the 30% PA6-reinforced PTFE composites have a significant improvement in mechanical performance. The improved tensile strength of PTFE composites is depicted in figure.

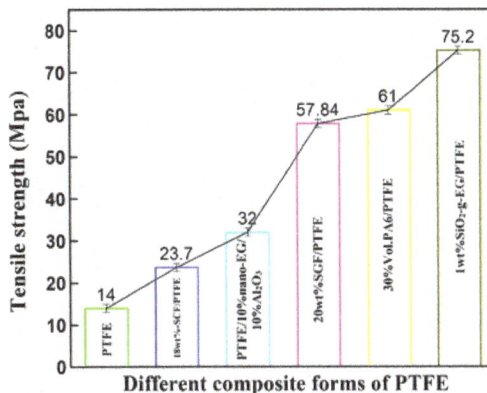

Figure: Tensile strength factor for various forms of PTFE composites.

## Chemical Properties of PTFE

The peculiar property of PTFE is chemical inertness. Naturally, PTFE is non-reactive and insoluble due to the strongly bonded carbon-fluorine atom. The high molecular weight is responsible for chemical inert behavior. PTFE is not affected by common reagents such as hydrofluoric, hydrochloric, and chlorosulfonic acids. Even above the transition temperature (327 °C), PTFE is insoluble in organic solvents like hydrocarbons, chlorinated hydrocarbons, or ester and phenol. This is due to the very fewer interaction forces between fluorocarbon and other molecules.

## Solubility of PTFE

The solvents chosen are oligomers, non-oligomeric perfluorocarbons, aromatic per-fluorocarbons, and non-perfluorocarbons. The report was consolidated the different types of thermodynamic solubility influence on PTFE. The solubility of PTFE involves various factors such as temperature, pressure, solvent polarity and swelling in solvents.

There are many practical issues of PTFE in terms of solubility. Several methods were employed to understand the solubility of PTFE with commercial solvents such as per-fluorocarbon and other halogenated fluids. Autogenous and superautogenous methods were involved in the solubility of PTFE under applied pressure. The report suggested that the entropy effects cause insolubility due to the less intermolecular forces. The molecular weight of the solvent can influence the solubility with the increase of lower critical solution temperature.

## Thermal Properties of PTFE

The performance in terms of thermal conductivity of PTFE over a wide range of tempera-ture is excellent than other polymers. The thermal stability is due to the linear high crystal-line arrangement of carbon-fluorine atoms that shows a high melting point of about 342 °C. For the measurement of crystallinity, different techniques can be preferred such as X-ray diffraction, density and dynamic mechanical analysis (DMA). The differential scanning cal-orimetry (DSC) technique was used to prepare the material from the melt with different crystallinity as a function of temperature. The sample was further tested with reference to one another. The thermal conductivity was measured using Lee's disk apparatus clearly indicates the improvement in heat transport of aluminum flakes included PTFE. The in-crease in thermal conductivity at 232 °C was noted for different levels of crystallinity. It was measured that the thermal conductivity of $Sr_2ZnSi_2O_7$ is 16.5 W/mK which is large when compared with PTFE (0.283 W/mK). The increase in thermal conductivity depends upon the filler material's shape, size, and thermal properties. The fillers generally provide the heat transfer path which was the reason for the increase in thermal conductivity.

## Thermal Transport Property of PTFE Composites

Thermal transport property of Al/PTFE nanocomposite with graphene and CNT were reported. Graphene and CNT are widely involving in numerous applications and sig-nificantly influence the material behavior which is added along with them. By intro-ducing graphene into Al/PTFE, increasing thermal conductivity was observed. Al acts as a mediator for heat transportation throughout the composite. Thermal diffusivity analysis of Al/PTFE portrayed about how quickly the material responds to the heat-ed environment. The addition of graphene in Al/PTFE increases thermal diffusivity in contrast to the addition of nano carbon (C) allotrope and CNT. The amorphous nature of nano C and CNT is due to the random arrangement of $sp^2$ and $sp^3$ carbons which re-sults in low thermo-physical property.

# Electrical Properties of PTFE

## Dielectric Property of PTFE

PTFE would play a role of a dielectric medium or insulating medium in an electronic component was consumed potentially because of distinct electric properties. The dielectric constant ($\varepsilon_r$) and the dissipation factor (tan$\delta$) are very important for a material operating as a dielectric medium in the charge storing devices. Depending upon the filler property, the $\varepsilon_r$ and tan$\delta$ varied and demonstrated in many reports. The improved $\varepsilon_r$ and tan$\delta$ for various PTFE-based composites are shown in Fig. It shows the PTFE composites tested under different frequency ranges and their respective $\varepsilon_r$ and tan$\delta$ values. It is obvious that depending upon the frequency, the polarization mechanism varies for different types of composites. For PTFE filled with $SiO_2$ (silicon dioxide), the values of $\varepsilon_r$ and tan$\delta$ increased at 5GHZ of frequency when compared to Virgin-PTFE. The large surface area of the $SiO_2$ and their moisture absorbance and contaminants were taken into account for explaining the function of $\varepsilon_r$ and tan$\delta$. PTFE/AlN (aluminum nitrate) showing improved $\varepsilon_r$ and tan$\delta$ as a function of filler loading. The values were obtained in the low frequency range from 100 Hz to 1 MHz which was suggested for electronic packaging. PTFE/$TeO_2$ showed excellent $\varepsilon_r$ and tan$\delta$ stability tested under 1 MHz and 7 GHz of the frequency range. The increase in tan$\delta$ was observed due to the interfacial polarization of the ceramic $TeO_2$ particles at higher volume fraction in the PTFE matrix. The experimental results showed the improved dielectric constant of $MgTiO_3$ ceramic filled PTFE. The results were good in agreement with the Maxwell-Garnett theoretical model which considers the occupation of ceramic particles in the host polymer system. The calcium copper titanate incorporated PTFE and its dielectric property was studied. The $\varepsilon_r$ here reported at low frequency (100 Hz) and attributed to interfacial polarization mechanism. The size of the particle present in the composites obviously changing the value of $\varepsilon_r$ and tan$\delta$ which were demonstrated. Over different frequency ranges, PTFE is stable and possess low dielectric constant $\varepsilon_r \sim 2.1$ and low loss tangent because of the neutralization of dipole moment exhibited by C-F bonds. A work was reported on the moisture absorbance of PTFE/Micron-rutile and PTFE/Nano-rutile composites. The moisture absorbing phenomena is important here because the water molecules are polar in nature having high $\varepsilon_r \sim 70$ which can significantly affect the dielectric nature of the PTFE composition.

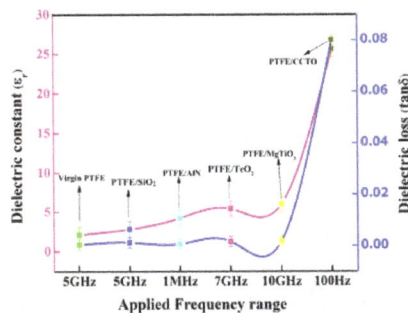

Figure: Reported dielectric constant ($\varepsilon r$) and dielectric loss (tan$\delta$) for PTFE composites tested under different frequency ranges.

## Optical and Spectral Properties of PTFE

The inherent optical and spectral properties of PTFE greatly help in the instrumentation of efficient optical devices. The light reflectance and diffusion parameters of PTFE are extremely high; hence, the material has been inevitable in optical instrumentation. Reflectance factor is the measurement of the surface's ability to reflect light which is equal to the ratio of reflected flux to the incident flux. PTFE exhibits good optical characteristics from a broad ultra-violet to near infra-red spectrum and good in performance when exposed to light or any other electromagnetic radiation. The reflectance angle measurements were studied using reflectometer which was used to measure the bidirectional reflectance of the PTFE pallet. The applications of PTFE as a light diffuser in radiometry were very attractive. The Lambertian surfaces (an ideal surface having high diffusive reflectance) are constructed with PTFE. Previous works considering that low density PTFE functions as a Lambertian diffuser. Measurement of bidirectional reflectance distribution function (BRDF), directional hemispherical reflectance (DHR) and directional hemispherical reflectance (DHT) were taken for two samples namely high density PTFE (HD PTFE) and low density PTFE (LD PTFE). To cover the entire wavelength of the spectrum, the aforesaid measurements were carefully done with the help of Fourier transform infrared Raman spectroscopy (FTIR) and LAMBDA 950 spectrophotometer. The results shown were in favor of LD PTFE because of the order of magnitude for DHR is less than HD PTFE.

The reflectance factor of PTFE is extreme to sustain at high intense electromagnetic radiation. For all optics-based instrumentation works, PTFE was suggested as a white light diffuser. A work was conducted to study the reflectance factor of pressed PTFE powder with a standard reflectance factor scale ratio ($45°/0°$). The sample was pressed and examined with $45°/0°$ reflectometer for wavelength varying from 380 to 770 nm. Analysis of samples was done by taking two variabilities: one is an operator (samples collected from 10 different laboratories) and another one is the material (various composition of PTFE). Final result evolved with the expanded uncertainty of $45°/0°$ reflectance factor due to material and operator variability.

Amorphous PTFE commercially known as Teflon®AF is having a glass-like transparency and possess good optical properties and highly preferred in optical devices. Teflon® AF is a copolymer of PDD and TFE. A detailed study was conducted to calculate the refractive index, extinction coefficient (k), the absorption coefficient ($\alpha$) and optical absorbance (A) of three different grades of Teflon®AF. The purpose of this work was to compare all the three grades for their respective optical properties.

## ECTFE

Ethylene Chlorotrifluoroethylene (ECTFE) is a type of fluoropolymer used as an electrostatic powder for coating applications on metallic surfaces. ECTFEs provide high performance and resistance to highly corrosive environments.

Ethylene Chlorotrifluoroethylene is often applied as a powder to protect metallic surfaces from corrosion. ECTFE possesses good film-forming characteristics and may be applied to metals without a primer, resulting in a high coating thickness. It provides increased permeability, tensile strength, and wear and creep resistance. Impact strength and other important corrosion resistance characteristics are present over a wide range of temperatures (-76°C to +150°C), making ECTFE popular in many industries.

Key features of ECTFE:

- Very good chemical and thermal resistance
- Optimum permeation resistance
- Outstanding flame resistance
- Very good surface characteristics
- Surface smoothness
- Purity

## ECTFE Properties

ECTFE is semi crystalline (50e60%) and melts at 240 C (commercial grade). It has an alpha relaxation at 140 C, a beta at 90 C, and gamma relaxation at 65 C. Conformation of ECTFE is an extended zigzag in which ethylene and CTFE alternate. The unit cell of ECTFE's crystal is hexagonal.

Similar to ETFE, ECTFE terpolymers (same termonomers) have better mechanical and abrasion and radiation resistance than PTFE and other perfluoropolymers.

Dielectric constant of ECTFE is 2.5e2.6, and independent of temperature and frequency. Dissipation factor is 0.02 and much larger than ETFE's.

ECTFE is resistant to most chemicals except hot polar and chlorinated solvents. It does not stress crack dissolve in any solvents. ECTFE has better barrier properties to $SO_2$, $Cl_2$, HCl, and water than FEP and PVDF.

## Applications

ECTFE can be applied in different ways:

- By electrostatic powder coating on metal surfaces.
- By rotolining on metal surfaces rotolining grade Halar 6012F.
- By sheet lining on metal surface or on fiber reinforced plastic FRP (glass fiber, carbon fiber).

- By extrusion or injection molding of self standing items, in particular pressure pipes.

- By rotomolding of self standing items like tanks or other shapes (rotomolding grade).

- As a protective film using an adequate adhesive.

ECTFE powder has been used for several decades in electrostatic powder coating. It has a good film forming characteristic, can be applied in certain cases directly to the metal without a primer and leads to a high coating thickness. ECTFE powder coatings have a typical thickness of 0,8 mm but can be applied up to 2 mm with a special grade for high build up.

Extrusion of ECTFE fabric backed sheets and subsequent fabrication to vessels, pipes or valves is used in the chemical industry. Typical sheet thickness is around 2,3 mm but can go up to 4 mm. Thick sheets are compression molded and can go up to 50 mm in thickness. They are used in the semiconductor industry for wet benches or for machining of other parts. Pressure piping systems composed of pipes, fittings and valves made of ECTFE are on the market up to 160 mm diameter. ECTFE rotomolded items complete the offer.

The most important application of ECTFE is in the field of corrosion protection. ECTFE has been used for several decades in different industries:

- Bleaching towers in pulp and paper.

- Sulphuric acid production and storage.

- Flue gas treatment in particular in the SNOX and WSA process.

- Electrolysis collectors or drying towers in the chlorine industry.

- Transport vessels for hazardous goods, in particular class 8.

- Halogen related industry (Bromine, chlorine, fluorine).

- Acid handling (sulphuric acid, nitric acid, phosphoric acid, hydrogen halides, hydrogen sulphide, etc).

- Mining applications, in particular high pressure heap leach.

ECTFE has been widely used in the semiconductor industry, for wet tool and tubing systems for lithographic chemicals.

ECTFE is also used in the pharmaceutical industry. ECTFE is useful when the equipment is exposed to chemical cleaning.

Moreover, ECTFE is used for primary and secondary jacketing in specialty cables like data cables or self-regulating heating cables, applications where good fire resistance and electrical properties are key properties. ECTFE is also used for braiding in that field.

ECTFE is available as monofilament fiber for applications in the chemical process industry and in flue gas treatment.

Latest developments in fibers include ECTFE nonwovens. ECTFE covers most of the aggressive chemicals which are covered by PTFE, but in contrast to PTFE, ECTFE fibers can be crimped. The nonwoven has low reactivity but high surface area and porosity. It is used as filters for highly reactive acids or alkalis even at elevated temperatures, coalescing of oil from water using the high surface energy or battery separators. Another application of ECTFE fibers is a so-called "Halar veil" that can be used to construct pipes or valves for use in corrosive environments such as chloralkali facilities.

Thanks to its good fire resistance and optical properties Halar films have been used for many years in the aircraft industry as well as for photovoltaic front and back sheets. For front sheets a UV blocking version is available. ECTFE Halar films have gained reputation as a lightweight front sheet in the Solar Impulse project of Bertrand Piccard, the first who aims to fly around the world in a solar driven plane.

ECTFE is used for manufacturing gaskets for storing liquid oxygen and other propellants for aerospace applications.

In many ways the chemical properties resemble PTFE but with a lower melting point.

## Perfluoroalkoxy Alkanes

Polytetrafluoroethylene (PTFE) is a synthetic fluoropolymer of tetrafluoroethylene that has numerous applications. The most widely known PTFE formulation is sold under the brand name of Teflon®. PTFE was discovered by DuPont Co. in 1938.

Perfluoroalkoxy alkanes (PFA) is a copolymer of hexafluoropropylene and perfluoroethers. It was developed after the discovery of PTFE by the same producer (DuPont Co.). One commonly known PFA formulation is Teflon PFA.

PFA has very similar properties to PTFE, though the biggest difference between PTFE and PFA is that PFA is melt-processed. This is accomplished through conventional injection molding as well as screw extrusion techniques.

### Area of Use

PTFE is popularly used as a non-stick coating for pans and many modern items of cookware. PTFE is often used in containers and pipes for handling reactive and corrosive chemicals. This is because it has non-reactive properties. Another practical application of PTFE is as a lubricant. Used in this way, PTFE helps to reduce friction within machinery, minimize the "wear and tear," and improve energy consumption.

PFA is generally used for plastic lab equipment because of its extreme resistance to chemical attack, optical transparency, and overall flexibility. PFA is also often used as tubing for handling critical or highly corrosive processes. Other applications for PFA are as sheet linings for chemical equipment. Because of its properties, it can facilitate the use of carbon steel fiber reinforced plastics (FRPs) as replacements for more expensive alloys and metals.

Devices used for level measurement are frequently exposed to harsh tank atmospheres. Aggressive media, such as acid, can corrode probes and parts in contact with the product. Viscous and sticky products are other level measurement challenges. These can cause build-up on the wetted parts of the transmitter and increase the risk of affecting the measurement. Wetted parts coated in PTFE or PFA materials provide a good resistance to corrosive products and are an effective solution for avoiding contamination caused by product build-up.

Water based products can cause condensation on the antenna, that might affect the level measurement. Due to the hydrophobic properties of PTFE and PFA, the measurement will remain unaffected in case droplets of water build-up on the antenna parts.

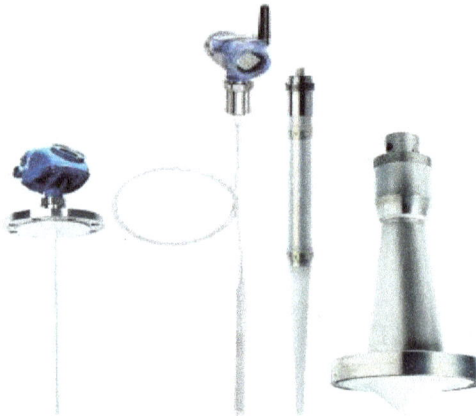

Figure: Level Measurement Devices Coated in PTFE or PFA Material

## Polyvinylidene Fluoride

PVDF is a tough, stable fluoropolymer with distinct engineering advantages.

Polyvinylidene fluoride (PVDF) is a synthetic resin produced by the polymerization of vinylidene fluoride ($CH_2=CF_2$). A tough plastic that is resistant to flame, electricity, and attack by most chemicals, PVDF is injection-molded into bottles for the chemical industry and extruded as a film for electrical insulation. Its flame resistance makes it especially desirable for insulating wire in buildings and aircraft. PVDF is also piezoelectric, changing its electrical charge in response to pressure and, conversely, exerting pressure in response to an applied electric field. This unique property makes it a good material for transducers in devices such as headphones, microphones, and sonic detectors.

## PVDF Properties

PVDF has a number of transitions and its density alters for each polymorph state. There are four known proposed states, named as $\alpha, \beta, \gamma,$ and . The most common phase is a-PVDF which exhibits transitions at 70 C ( $\gamma$ ), 38 C( $\beta$ ), 50 C ( $\alpha''$ ), and 100 C ( $\alpha'$ ).

PVDF resists most organic and inorganic chemicals including chlorinated solvents. Strong bases, amines, esters, and ketones attack this resin. The impact ranges from swelling to complete dissolution in these solvents depending on the conditions. PVDF exhibits compatibility with a number of polymers. Commercially useful blends with acrylics and methacrylics have been developed.

PVDF, just as ETFE, readily crosslinks as a result of exposure to radiation. Radiation (gamma rays) has modest effect on the mechanical properties of PVDF.

## Polychlorotrifluoroethylene

Polychlorotrifluoroethylene (PCTFE) is a synthetic resin formed by the polymerization of chlorotrifluoroethylene.

It is a moldable, temperature-resistant, and chemical-resistant plastic that finds specialty applications in the chemical, electrical, and aerospace industries.

PCTFE can be prepared as a powder by treating an aqueous suspension or emulsion of chlorotrifluoroethylene with polymerization catalysts. The repeating units of the polymeric moleculehave the following structure:

$$[CF_2 - CFCI] \cdot$$

PCTFE powder can be melted and then shaped by molding or by extrusion into solid. The plastic remains ductile at temperatures as low as −200 °C (−330 °F) and is stable at temperatures above 200 °C (390 °F). It resists attack by most chemicals, is impermeable to gases, retains its properties upon exposure to gamma radiation, and is an excellent electrical insulator. Because of these properties, it is employed in seals, gaskets, and barriers for cryogenic (ultralow-temperature), petrochemical, aerospace, and uranium-enrichment equipment.

## PCTFE Properties

PCTFE is a semicrystalline polymer with a helical polymer chain and a pseudohexagonal crystal. Crystal growth is spherulitic and consists of folded chains. Large size of chlorine constrains recrystallization after melting during processing. This resin has good properties at cryogenic temperatures relative to plastics in general, although they are inferior to other fluoropolymers except PVDF.

PCTFE has exceptional barrier properties and superb chemical resistance. It is attacked by a number of organic solvents.

PCTFE has low thermal stability and degrades upon reaching its melting point, requiring special care during processing.

### References

- Polymer Handbook, J. Brandrup (Editor), E. H. Immergut (Editor), E. A. Grulke (Editor), Publication Date: May 29, 2003 ISBN 978-0471479369

- What-is-fluoropolymer-coating: toefco.com, Retrieved 10 July 2018

- Ethylene-chlorotrifluoroethylene-ectfe-4849: corrosionpedia.com, Retrieved 19 May 2018

- White-paper-ptfe-pfa-similarities-differences-en-585104: emerson.com, Retrieved 31 March 2018

- Polyvinylidene-fluoride, science: britannica.com, Retrieved 17 June 2018

# Polymerization

Polymerization is the process of forming polymer chains or three-dimensional networks by reacting monomer molecules together in a chemical reaction. All the different types of polymerization such as radical polymerization, step-growth polymerization and chain-growth polymerization, etc. have been carefully analyzed in this chapter.

Polymerization is a process in which relatively small molecules, called monomers, combine chemically to produce a very large chainlike or network molecule, called a polymer.

The monomer molecules may be all alike, or they may represent two, three, or more different compounds. Usually at least 100 monomer molecules must be combined to make a product that has certain unique physical properties—such as elasticity, high tensile strength, or the ability to form fibres—that differentiate polymers from substances composed of smaller and simpler molecules; often, many thousands of monomer units are incorporated in a single molecule of a polymer. The formation of stable covalent chemical bonds between the monomers sets polymerization apart from other processes, such as crystallization, in which large numbers of molecules aggregate under the influence of weak intermolecular forces.

Two classes of polymerization usually are distinguished. In condensation polymerization, each step of the process is accompanied by the formation of a molecule of some simple compound, often water. In addition polymerization, monomers react to form a polymer without the formation of by-products. Addition polymerizations usually are carried out in the presence of catalysts, which in certain cases exert control over structural details that have important effects on the properties of the polymer.

Functional group: monomers and polymers Functional groups in monomers and polymers.

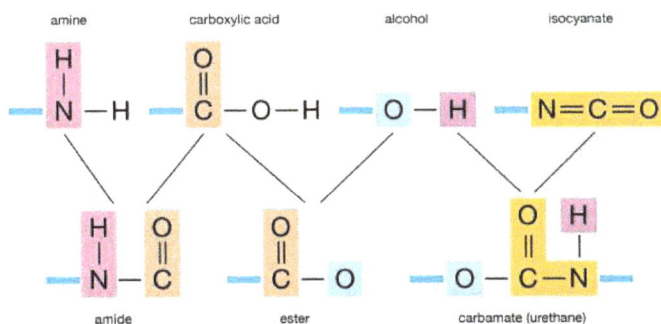

Linear polymers, which are composed of chainlike molecules, may be viscous liquids or solids with varying degrees of crystallinity; a number of them can be dissolved in certain liquids, and they soften or melt upon heating. Cross-linked polymers, in which the molecular structure is a network, are thermosetting resins (i.e., they form under the influence of heat but, once formed, do not melt or soften upon reheating) that do not dissolve in solvents. Both linear and cross-linked polymers can be made by either addition or condensation polymerization.

## Classification

The polymerization reactions are classified as:

- Radical Polymerization

- Ionic Polymerization:

    o   Anionic polymerization

    o   Cationic polymerization

## Radical Polymerization

In order to produce radical polymerization it´s necessary that initiator monomer or hardener, which activates and causes the reaction, contains free radicals, namely unpaired electrons which react with the resin monomer in order to form the polymer.

Radical is defined as chemical material extremely unstable therefore with a high reactive power due it has unpaired electrons.

The steps that occur in the radical polymerization are:

Initiation of the reaction - free radical is formed by the action of chemical, thermal, electrochemical or photochemical energy.

- Formation of radical-R --- RAD> RAD * + * R

- Chain initiation RAD * + A ---> RAD-A *

Chain growth

- RAD-A * + A ---> RAD-A-A *

- RAD-A-A * + A ---> RAD-A-A-A *

Chain termination - There are several ways to end the reaction by either:

- Combine 2 radicals RAD-AAA-AAA * + * --- RAD> RAD-AAAAAA-RAD

- Using inhibitors / regulators – external agents of the polymerization which react with the radical making terminate the reaction.

One of the main disadvantages of radical polymerization is that it can not control the molecular weight and size of the final polymer, because each reaction terminates in an undefined way.

The polymers that are produced by radical polymerization are strongly temperature dependent, if temperature increase will speed up the polymerization reactions causing:

- Shorter chains of the polymers and decreasing their mechanical properties (strength and elongation).

- Reduction of working time, pot life and curing.

## Ionic Polymerization

Ionic polymerization operation is similar to the radical polymerization; the radical in this case is an ion (atom or molecule) in which one of its parts is charged positively or negatively by the absence or presence of electrons.

Features:

- Requires less activation energy than the radical polymerization.

- It is not as temperature dependent.

- The chain termination occurs only for the use of inhibitors, regulators or other agents that stop the reaction.

Depending on the charge of the ion, ionic polymerization is classified into:

- Anionic polymerization

- Cationic Polymerization

Anionic polymerization - The ion is an atom or molecule with excess of electrons - negatively charged.

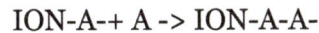

ION-+ A -> ION-A-

ION-A-+ A -> ION-A-A-

ION-A-A-+ A -> ION-A-A-A-

Cationic Polymerization - The ion is an atom or molecule with lack of electrons - positively charged.

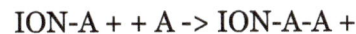

ION + + A -> ION-A +

ION-A + + A -> ION-A-A +

ION-A-A + + A -> ION-A-A-A +

## Adhesives Cure by Polymerization:

- Radical polymerization

    o MMA

    o Anaerobic

    o Unsaturated polyesters

    o Acrylic curing by ultraviolet.

- Ionic Polymerization:

    o Anionic - Cyanoacrylates

    o Cationic - Radiation curing Epoxy

# Radical Polymerization

Radical polymerization belongs to the most important reactions running by chain mechanism. Mechanism of polymer chain creation can be described by elementary actions: initiation, propagation, termination and transfer reactions. Succession of actions running at construction of one macromolecule is very quick (around one second). However, great number of macromolecules is produced simultaneously in the system, which is why all types of partial reaction occur simultaneously.

## Initiation

The initiation reaction produces free radicals. There are several ways to do this:

- Chemical initiation The decomposition of the initiator (e.g. AIBN) forms free radicals:

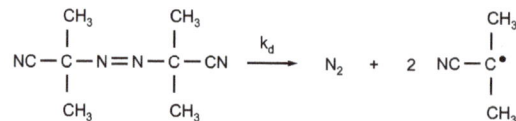

$$NC-\underset{\underset{CH_3}{|}}{\overset{\overset{CH_3}{|}}{C}}-N=N-\underset{\underset{CH_3}{|}}{\overset{\overset{CH_3}{|}}{C}}-CN \xrightarrow{k_d} N_2 \;+\; 2 \; NC-\underset{\underset{CH_3}{|}}{\overset{\overset{CH_3}{|}}{C}}{}^{\bullet}$$

$$I_2 \xrightarrow{k_d} I^{\bullet} + I^{\bullet} \quad (r_d = k_d I_2)$$

$$I^{\bullet} + M \xrightarrow{k_I} R_1^{\bullet}$$

$$\frac{dI^{\bullet}}{dt} = 2f\,k_d I_2 - M \approx 0$$

$$\Rightarrow k_I I^{\bullet} M = 2f k_d I_2 \equiv R_I$$

where $f$ is the initiator efficiency, typically $f = [0.5, 1]$. Note that in order to ensure a continuous production of radicals all over the process, $1/k_d$ should be larger than the characteristic time of the polymerization reaction. Examples of the decomposition characteristic time, $\tau_d$ for some commercial initiators are:

|  | $\tau_d$ | $T$ |
|---|---|---|
| Acetyl peroxide | 2 h | $80^\circ\,C$ |
| Cumyl peroxide | 12 h | $110^\circ\,C$ |
| t-Butyl hydroperoxide | 45 h | $150^\circ\,C$ |

Since this is a first order process, $\tau_d = 1/k_d$.

1. Thermal Initiation: Thermal Decomposition of the Monomer (e.g. styrene)

    This represents a danger, for example during monomer transportation, since it may lead to undesired polymerization of the monomer. For this reason, inhibitors (scavengers of radicals) are usually added to the monomers before storage. This causes the occurrence of a non reproducible induction period when such monomers are polymerized.

2. Initiation by Radiation

    The decomposition of the initiator is caused by light or another source of radiation. Since this method is quite expensive, it is only applied to polymerization systems operating at very low temperatures.

## Propagation

Propagation is the addition of a monomer molecule to a radical chain.

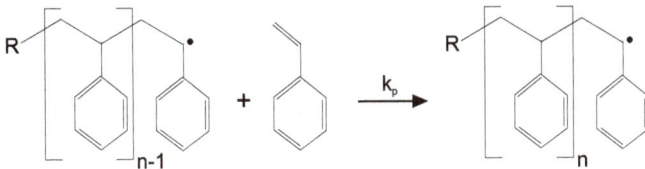

$$R_n^\bullet + M \xrightarrow{k_d} R_{n+1}^\bullet \quad \left( r_p = k_p\, R_n^\bullet\, M \right)$$

## Chain Transfer

- Chain transfer to monomer

$$R_n^\bullet + M \xrightarrow{k_{fm}} P_n + R_1^\bullet \quad \left( r = k_{fm}\, R_n^\bullet\, M \right)$$

The reactants are the same as for the propagation reaction, but the activation energy is much larger. Accordingly, $k_{fm}$ is usually at least 103 times smaller than $k_p$. This reaction

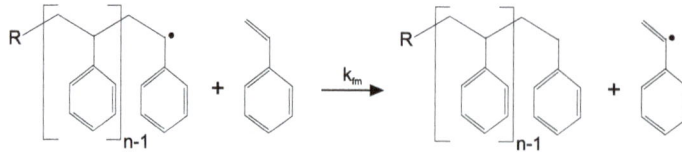

leads to the formation of a polymer chain with a terminal double bond. This can induce chain branching through the terminal double bond propagation reaction.

- Chain transfer to chain transfer agent

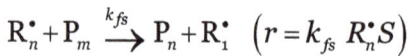

$$R_n^{\bullet}+P_m \xrightarrow{k_{fs}} P_n+R_1^{\bullet} \quad \left(r=k_{fs}\, R_n^{\bullet}S\right)$$

A chain transfer agent, S is a molecule containing a weak bond that can be broken to lead to radical transfer, similarly as in the case of monomer above (e.g. $CCl_4$, $CBr_4$, mercaptans).

- Chain transfer to polymer

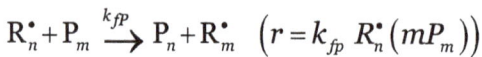

$$R_n^{\bullet}+P_m \xrightarrow{k_{fp}} P_n+R_m^{\bullet} \quad \left(r=k_{fp}\, R_n^{\bullet}\left(mP_m\right)\right)$$

In this reaction the growing radical chain, $R_n^{\bullet}$ extracts a hydrogen from the dead chain, $P_m$. Since this extraction can occur on any of the m monomer units along the chain, the rate of this reaction is proportional to the length of $P_m$.

General observations on the role of chain transfer reactions:

- The concentration of radicals is not affected and therefore the rate of monomer consumption is also unchanged.

- The growth of polymer chains is stopped and therefore shorter chains are produced.

- Each transfer event leaves a different end-group on the chain that can be detected (NMR, titration) so as to identify and quantify the corresponding chain transfer reaction.

- Nonlinear (branched) polymer chains are produced: directly by chain transfer to polymer or indirectly through the propagation of the terminal double bond left by a chain transfer to monomer event.

## Bimolecular Termination

Bimolecular termination occurs according to two different mechanisms: termination by combination and termination by disproportionation. Their relative importance depends upon the specific polymerization system.

- Termination by combination

$$R_n^{\bullet}+R_m^{\bullet} \xrightarrow{k_{tc}} P_{n+m} \quad \left(r = k_{tc}\, R_n^{\bullet}\, R_m^{\bullet}\right)$$

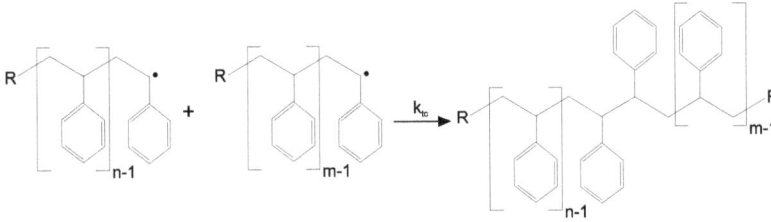

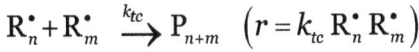

This reaction results in an increase of the chain length.

- Termination by disproportionation

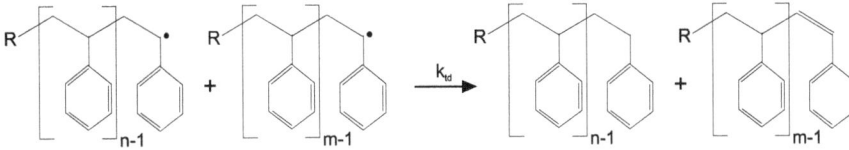

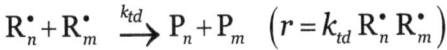

$$R_n^{\bullet}+R_m^{\bullet} \xrightarrow{k_{td}} P_n + P_m \quad \left(r = k_{td}\, R_n^{\bullet}\, R_m^{\bullet}\right)$$

The chain length remains constant during the termination reaction. Note that also in this case chains with terminal double bond are produced, which can therefore lead to the occurrence of branching.

## Methods

There are four industrial methods of radical polymerization:

- *Bulk polymerization:* reaction mixture contains only initiator and monomer, no solvent.

- *Solution polymerization:* reaction mixture contains solvent, initiator, and monomer.

- *Suspension polymerization:* reaction mixture contains an aqueous phase, water-insoluble monomer, and initiator soluble in the monomer droplets (both the monomer and the initiator are hydrophobic).

- *Emulsion polymerization:* similar to suspension polymerization except that the initiator is soluble in the aqueous phase rather than in the monomer droplets (the monomer is hydrophobic, and the initiator is hydrophilic). An emulsifying agent is also needed.

Other methods of radical polymerization include the following:

- *Template polymerization*: In this process, polymer chains are allowed to grow along template macromolecules for the greater part of their lifetime. A well-chosen template can affect the rate of polymerization as well as the molar mass and microstructure of the daughter polymer. The molar mass of a daughter polymer can be up to 70 times greater than those of polymers produced in the absence of the template and can be higher in molar mass than the templates themselves. This is because of retardation of the termination for template-associated radicals and by hopping of a radical to the neighboring template after reaching the end of a template polymer.

- *Plasma polymerization*: The polymerization is initiated with plasma. A variety of organic molecules including alkenes, alkynes, and alkanes undergo polymerization to high molecular weight products under these conditions. The propagation mechanisms appear to involve both ionic and radical species. Plasma polymerization offers a potentially unique method of forming thin polymer films for uses such as thin-film capacitors, antireflection coatings, and various types of thin membranes.

- *Sonication*: The polymerization is initiated by high-intensity ultrasound. Polymerization to high molecular weight polymer is observed but the conversions are low (<15%). The polymerization is self-limiting because of the high viscosity produced even at low conversion. High viscosity hinders cavitation and radical production.

## Reversible Deactivation Radical Polymerization

Also known as living radical polymerization, controlled radical polymerization, reversible deactivation radical polymerization (RDRP) relies on completely pure reactions, preventing termination caused by impurities. Because these polymerizations stop only when there is no more monomer, polymerization can continue upon the addition of more monomer. Block copolymers can be made this way. RDRP allows for control of molecular weight and dispersity. However, this is very difficult to achieve and instead a pseudo-living polymerization occurs with only partial control of molecular weight and dispersity. ATRP and RAFT are the main types of complete radical polymerization.

- *Atom Transfer Radical Polymerization (ATRP):* based on the formation of a carbon-carbon bond by atom transfer radical addition. This method, independently discovered in 1995 by Mitsuo Sawamoto and by Jin-Shan Wang and Krzysztof Matyjaszewski, requires reversible activation of a dormant species (such as an alkyl halide) and a transition metal halide catalyst (to activate dormant species).

- *Reversible Addition-Fragmentation Chain Transfer Polymerization (RAFT):* requires a compound that can act as a reversible chain transfer agent, such as dithio compounds.

- *Stable Free Radical Polymerization (SFRP)*: used to synthesize linear or branched polymers with narrow molecular weight distributions and reactive end groups on each polymer chain. The process has also been used to create block co-polymers with unique properties. Conversion rates are about 100% using this process but require temperatures of about 135 °C. This process is most commonly used with acrylates, styrenes, and dienes. The reaction scheme in Figure illustrates the SFRP process.

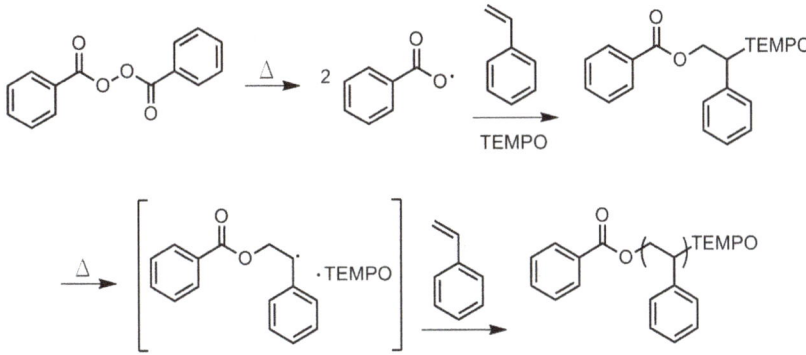

*Figure*: Reaction scheme for SFRP.

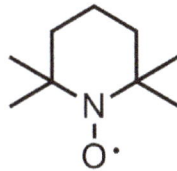

*Figure*: TEMPO molecule used to functionalize the chain ends.

Because the chain end is functionalized with the TEMPO molecule Figure, premature termination by coupling is reduced. As with all living polymerizations, the polymer chain grows until all of the monomer is consumed.

## Kinetics

In typical chain growth polymerizations, the reaction rates for initiation, propagation and termination can be described as follows:

$$v_i = \mathrm{d}[M\cdot]/\mathrm{d}t = 2k_d f[I]$$

$$v_p = k_p[M][M\cdot]$$

$$v_t = -\mathrm{d}[M\cdot]/\mathrm{d}t = 2k_t[M\cdot]^2$$

where $f$ is the efficiency of the initiator and $k_d$, $k_p$, and $k_t$ are the constants for initiator dissociation, chain propagation and termination, respectively. [I] [M] and [M•] are the concentrations of the initiator, monomer and the active growing chain.

Under the steady state approximation, the concentration of the active growing chains remains constant, i.e. the rates of initiation and of termination are equal. The concentration of active chain can be derived and expressed in terms of the other known species in the system,

$$[M \cdot] = \left( \frac{k_d [I] f}{k_t} \right)^{1/2}$$

In this case, the rate of chain propagation can be further described using a function of the initiator and monomer concentrations,

$$v_p = k_p \left( \frac{f k_d}{k_t} \right)^{1/2} [I]^{1/2} [M]$$

The kinetic chain length v is a measure of the average number of monomer units reacting with an active center during its lifetime and is related to the molecular weight through the mechanism of the termination. Without chain transfer, the kinetic chain length is only a function of propagation rate and initiation rate,

$$v = \frac{v_p}{v_i} = \frac{k_p [M][M \cdot]}{2 f k_d [I]} = \frac{k_p [M]}{2 (f k_d k_t [I])^{1/2}}$$

Assuming no chain transfer effect occurs in the reaction, the number average degree of polymerization $P_n$ can be correlated with the kinetic chain length. In the case of termination by disproportionation, one polymer molecule is produced per every kinetic chain:

$$x_n = v$$

Termination by combination leads to one polymer molecule per two kinetic chains:

$$x_n = 2v$$

Any mixture of both these mechanisms can be described by using the value $\delta$, the contribution of disproportionation to the overall termination process:

$$x_n = \frac{2}{1 + \delta} v$$

If chain transfer is considered, the kinetic chain length is not affected by the transfer process because the growing free-radical center generated by the initiation step stays alive after any chain transfer event, although multiple polymer chains are produced. However, the number average degree of polymerization decreases as the chain transfers, since the growing chains are terminated by the chain transfer events. Taking into account the chain transfer reaction towards solvent $S$, initiator $I$, polymer $P$, and added chain transfer agent $T$. The equation of $P_n$ will be modified as follows:

$$\frac{1}{x_n} = \frac{2k_{t,d}+k_{t,c}}{k_p^2[M]^2}v_p + C_M + C_S\frac{[S]}{[M]} + C_I\frac{[I]}{[M]} + C_P\frac{[P]}{[M]} + C_T\frac{[T]}{[M]}$$

It is usual to define chain transfer constants C for the different molecules

$$C_M = \frac{k_{tr}^M}{k_p}, \quad C_S = \frac{k_{tr}^S}{k_p}, \quad C_I = \frac{k_{tr}^I}{k_p}, \quad C_P = \frac{k_{tr}^P}{k_p}, \quad C_T = \frac{k_{tr}^T}{k_p},$$

## Thermodynamics

In chain growth polymerization, the position of the equilibrium between polymer and monomers can be determined by the thermodynamics of the polymerization. The Gibbs free energy ($\Delta G_p$) of the polymerization is commonly used to quantify the tendency of a polymeric reaction. The polymerization will be favored if $\Delta G_p < 0$; if $\Delta G_p > 0$, the polymer will undergo depolymerization. According to the thermodynamic equation $\Delta G = \Delta H - T\Delta S$, a negative enthalpy and an increasing entropy will shift the equilibrium towards polymerization.

In general, the polymerization is an exothermic process, i.e. negative enthalpy change, since addition of a monomer to the growing polymer chain involves the conversion of π bonds into σ bonds, or a ring–opening reaction that releases the ring tension in a cyclic monomer. Meanwhile, during polymerization, a large amount of small molecules are associated, losing rotation and translational degrees of freedom. As a result, the entropy decreases in the system, $\Delta S_p < 0$ for nearly all polymerization processes. Since depolymerization is almost always entropically favored, the $\Delta H_p$ must then be sufficiently negative to compensate for the unfavorable entropic term. Only then will polymerization be thermodynamically favored by the resulting negative $\Delta G_p$.

In practice, polymerization is favored at low temperatures: $T\Delta S_p$ is small. Depolymerization is favored at high temperatures: $T\Delta S_p$ is large. As the temperature increases, $\Delta G_p$ become less negative. At a certain temperature, the polymerization reaches equilibrium (rate of polymerization = rate of depolymerization). This temperature is called the ceiling temperature ($T_c$). $\Delta G_p = 0$.

## Stereochemistry

The stereochemistry of polymerization is concerned with the difference in atom connectivity and spatial orientation in polymers that has the same chemical composition. Staudinger studied the stereoisomerism in chain polymerization of vinyl monomers in late 1920s, and it took another two decades for people to fully appreciate the idea that each of the propagation steps in the polymer growth could give rise to stereoisomerism.

The major milestone in the stereochemistry was established by Ziegler and Natta and their coworkers in 1950s, as they developed metal based catalyst to synthesize stereo-regular polymers. The reason why the stereochemistry of the polymer is of particular interest is because the physical behavior of a polymer depends not only on the general chemical composition but also on the more subtle differences in microstructure. Atactic polymers consist of a random arrangement of stereochemistry and are amorphous (noncrystalline), soft materials with lower physical strength. The corresponding isotactic (like substituents all on the same side) and syndiotactic (like substituents of alternate repeating units on the same side) polymers are usually obtained as highly crystalline materials. It is easier for the stereoregular polymers to pack into a crystal lattice since they are more ordered and the resulting crystallinity leads to higher physical strength and increased solvent and chemical resistance as well as differences in other properties that depend on crystallinity. The prime example of the industrial utility of stereoregular polymers is polypropene. Isotactic polypropene is a high-melting (165 °C), strong, crystalline polymer, which is used as both a plastic and fiber. Atactic polypropene is an amorphous material with an oily to waxy soft appearance that finds use in asphalt blends and formulations for lubricants, sealants, and adhesives, but the volumes are minuscule compared to that of isotactic polypropene.

When a monomer adds to a radical chain end, there are two factors to consider regarding its stereochemistry: 1) the interaction between the terminal chain carbon and the approaching monomer molecule and 2) the configuration of the penultimate repeating unit in the polymer chain. The terminal carbon atom has $sp^2$ hybridization and is planar. Consider the polymerization of the monomer $CH_2=CXY$. There are two ways that a monomer molecule can approach the terminal carbon: the mirror approach (with like substituents on the same side) or the non-mirror approach (like substituents on opposite sides). If free rotation does not occur before the next monomer adds, the mirror approach will always lead to an isotactic polymer and the non-mirror approach will always lead to a syndiotactic polymer (Figure).

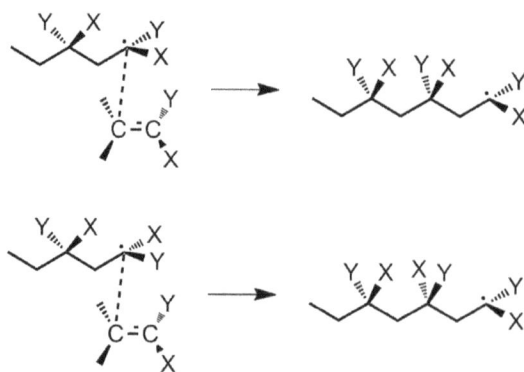

*Figure:* (Top) formation of isotactic polymer; (bottom) formation of syndiotactic polymer.

However, if interactions between the substituents of the penultimate repeating unit and the terminal carbon atom are significant, then conformational factors could cause the monomer to add to the polymer in a way that minimizes steric or electrostatic interaction Figure.

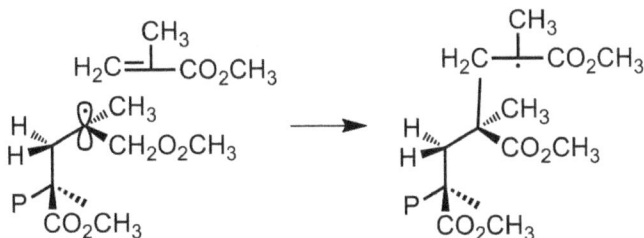

Figure: Penultimate unit interactions cause monomer to add in a way that minimizes steric hindrance between substituent groups. (P represents polymer chain.)

## Reactivity

Traditionally, the reactivity of monomers and radicals are assessed by the means of copolymerization data. The $Q$–$e$ scheme, the most widely used tool for the semi-quantitative prediction of monomer reactivity ratios, was first proposed by Alfrey and Price in 1947. The scheme takes into account the intrinsic thermodynamic stability and polar effects in the transition state. A given radical $M_i^o$ and a monomer $M_j$ are considered to have intrinsic reactivities $P_i$ and $Q_j$, respectively. The polar effects in the transition state, the supposed permanent electric charge carried by that entity (radical or molecule), is quantified by the factor $e$, which is a constant for a given monomer, and has the same value for the radical derived from that specific monomer. For addition of monomer 2 to a growing polymer chain whose active end is the radical of monomer 1, the rate constant, $k_{12}$, is postulated to be related to the four relevant reactivity parameters by

$$k_{12} = P_1 Q_2 \exp(-e_1 e_2)$$

The monomer reactivity ratio for the addition of monomers 1 and 2 to this chain is given by

$$r_1 = \frac{k_{11}}{k_{12}} = \frac{Q_1}{Q_2} \exp(-e_1(e_1 - e_2))$$

For the copolymerization of a given pair of monomers, the two experimental reactivity ratios $r_1$ and $r_2$ permit the evaluation of $(Q_1/Q_2)$ and $(e_1 - e_2)$. Values for each monomer can then be assigned relative to a reference monomer, usually chosen as styrene with the arbitrary values $Q = 1.0$ and $e = -0.8$.

## Applications

Free radical polymerization has found applications including the manufacture of polystyrene, thermoplastic block copolymer elastomers, cardiovascular stents, chemical surfactants and lubricants. Block copolymers are used for a wide variety of applications including adhesives, footwear and toys.

Free radical polymerization has uses in research as well, such as in the functionalization of carbon nanotubes. CNTs intrinsic electronic properties lead them to form large aggregates in solution, precluding useful applications. Adding small chemical groups to the walls of CNT can eliminate this propensity and tune the response to the surrounding environment. The use of polymers instead of smaller molecules can modify CNT properties (and conversely, nanotubes can modify polymer mechanical and electronic properties). For example, researchers coated carbon nanotubes with polystyrene by first polymerizing polystyrene via chain radical polymerization and subsequently mixing it at 130 °C with carbon nanotubes to generate radicals and graft them onto the walls of carbon nanotubes Figure. Chain growth polymerization ("grafting to") synthesizes a polymer with predetermined properties. Purification of the polymer can be used to obtain a more uniform length distribution before grafting. Conversely, "grafting from", with radical polymerization techniques such as atom transfer radical polymerization (ATRP) or nitroxide-mediated polymerization (NMP), allows rapid growth of high molecular weight polymers.

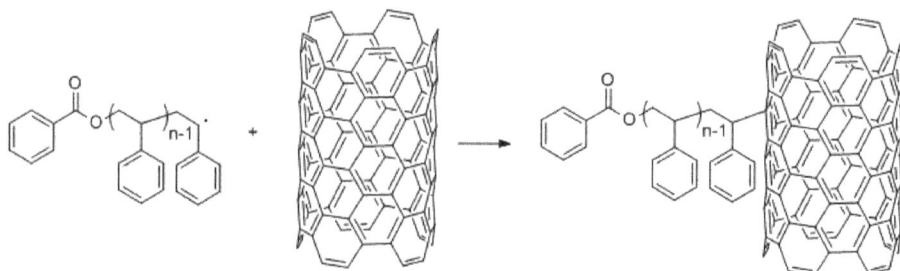

*Figure*: Grafting of a polystyrene free radical onto a single-walled carbon nanotube.

Radical polymerization also aids synthesis of nanocomposite hydrogels. These gels are made of water-swellable nano-scale clay (especially those classed as smectites) enveloped by a network polymer. They are often biocompatible and have mechanical properties (such as flexibility and strength) that promise applications such as synthetic tissue. Synthesis involves free radical polymerization. The general synthesis procedure is depicted in Figure. Clay is dispersed in water, where it forms very small, porous plates. Next the initiator and a catalyst are added, followed by adding the organic monomer, generally an acrylamide or acrylamide derivative. The initiator is chosen to have stronger interaction with the clay than the organic monomer, so it preferentially adsorbs to the clay surface. The mixture and water solvent is heated to initiate polymerization. Polymers grow off the initiators that are in turn bound to the clay. Due to recombination and disproportionation reactions, growing polymer chains bind to one another, forming a strong, cross-linked network polymer, with clay particles acting as branching points for multiple polymer chain segments. Free radical polymerization used in this context allows the synthesis of polymers from a wide variety of substrates (the chemistries of suitable clays vary). Termination reactions unique to chain growth polymerization produce a material with flexibility, mechanical strength and biocompatibility.

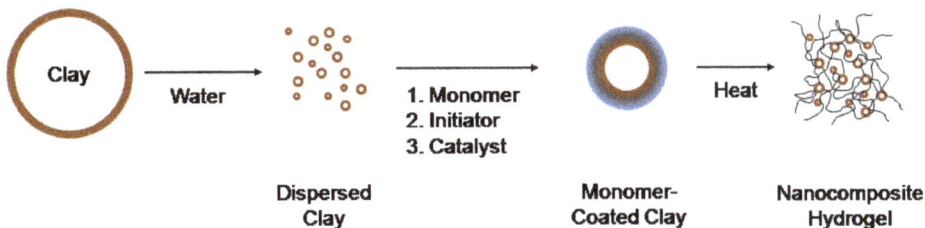

*Figure*: General synthesis procedure for a nanocomposite hydrogel.

## Electronics

The radical polymer glass PTMA is about 10 times more electrically conductive than common semiconducting polymers. PTMA is in a class of electrically active polymers that could find use in transparent solar cells, antistatic and antiglare coatings for mobile phone displays, antistatic coverings for aircraft to protect against lightning strikes, flexible flash drives, and thermoelectric devices, which convert electricity into heat and the reverse. Widespread practical applications require increasing conductivity another 100 to 1,000 times.

The polymer was created using deprotection, which involves replacing a specific hydrogen atom in the pendant group with an oxygen atom. The resulting oxygen atom in PTMA has one unpaired electron in its outer shell, making it amenable to transporting charge. The deprotection step can lead to four distinct chemical functionalities, two of which are promising for increasing conductivity.

## Emulsion Polymerization

Emulsion polymerization is a type of radical polymerization. Emulsion polymerization is one of the most important methods for the polymerization of a large number of monomers, like vinyl acetate, vinyl chloride, chloroprene, acrylamide, acrylates, and methacrylates. It is also used for the production of various copolymers, like acrylonitrile-butadiene-styrene (ABS).

Emulsion polymerization has several advantages over other polymerization techniques; for example, it is more rapid than bulk or solution polymerization at the same temperature, the conversion is essentially 100 percent, and the average molecular weight is usually (much) higher than at the same polymerization rate in bulk or solution polymerization. Also, heat dissipation and viscosity control are much less problematic than in bulk polymerization.

In general, an emulsion polymerization system consists of a dispersing medium, monomer, emulsifier, initiator and, if necessary, modifiers. Water[1] is normally the continuous phase in which the various components are dispersed by the emulsifiers. The monomers are only slightly soluble in water. They form droplets that are suspended and stabilized by the emulsifiers, that is, the emulsifier molecules associate and form micelles that surround small amounts of monomer. The remaining monomer is dispersed in small droplets.

Common emulsifiers are anionic and nonionic surfactants. A combination of both types will often improve the stability of the dispersed droplets. Typical anionic emulsifiers are sodium, potassium, or ammonium salts of fatty acids and $C_{12}$ - $C_{16}$ alkyl sulfates. Typical nonionic surfactants are poly(ethylene oxide), poly(vinyl alcohol) and hydroxyethyl cellulose.

The first hypothesis of the mechanism of emulsion polymerization was proposed by Harkins[2]; The process can be divided in three distinctive stages:

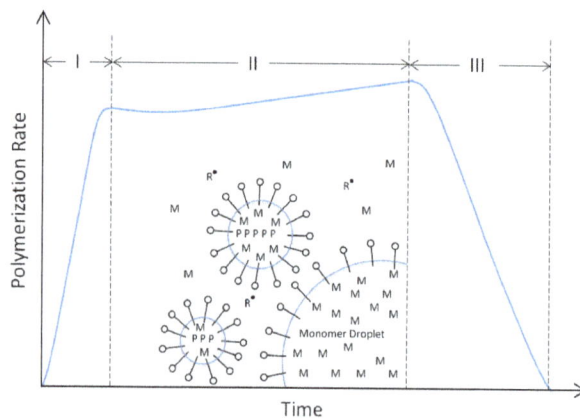

Stages of an emulsion polymerization

# STAGE I

At the beginning, the mixture consists solely of the continuous water phase with dispersed surfactant micelles[3] and emulsified monomer droplets of 1-10 microns. Most monomer is localized in these droplets and some is dissolved in the micelles and a little in the water phase. The micelles are in a dynamic equilibrium with the dissolved emulsifier molecules. Nucleation stops when the surface area becomes large enough to absorb all of the emulsifier molecules. The polymerization starts when initiator is added. A common initiator is (water soluble) potassium persulfate. It decomposes in the water and forms negatively charged sulfate radicals:

$$S_2O_8^- + \text{Heat} \rightarrow 2\ SO_{4\cdot}^-$$

These radicals react with the in the water dissolved monomers and form soap-type free radicals.

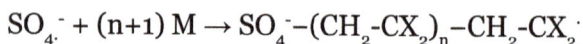

$$SO_{4\cdot}^- + (n+1)\ M \rightarrow SO_4^--(CH_2-CX_2)_n-CH_2-CX_2^\cdot$$

They either associate with the dissolved emulsifier molecules to micelles or they migrate into existing micelle droplets. During the first stage of the polymerization, monomers continuously migrate from the large monomer droplets through the water phase into the micelles and are added to the growing polymer chains. At the same time, new particles are formed by the initiators. Some polymerization takes also place in the water

phase due to the solubility of the monomers and initiators, whereas the monomer droplets do not provide loci for polymerization because the negatively charged surfactant molecules that surround these droplets are virtually impossible to penetrate by the also negatively charged initiator molecules.

## STAGE II

The number of polymer particles and the rate of polymerization increase as long as new radicals and polymer micelles are formed. Eventually, all surfactant in the system has been absorbed or all initiator molecules have been used up by the polymer particles. At this point in time, the rate of polymerization remains more or less constant. The particle number stabilizes at a rather low value which is only a small fraction, typically about 0.1% of the number of micelles initially present.

The diffusion of the free radicals into the monomer droplets is generally not important to the reaction rate during the growth stage and the rate of monomer diffusion is usually adequate to supply enough monomer to keep the reaction in the particles going.

## STAGE III

As the size of the latex particles increases, the size of the monomer droplets decreases and eventually they disappear. At this stage, the reaction mixture consists solely of monomer swollen polymer particles, the so-called latex particles, and dissolved monomers. Since monomer droplets are no longer present, both the monomer concentration and the reaction rate steadily decrease with time. With increasing polymer concentration, some cross-linking reactions and branching can also be expected. The reaction ends when either all monomers are used up or when a second radical diffuses into the polymer particles and causes immediate bimolecular termination by disproportionation or recombination of two radicals. If no termination occurs, the system reaches a conversion of essentially 100 percent. The final (submicroscopic) polymer particles are spherical in shape and have a diameter in the range of 50 – 300 nm which is between the initial micelle and monomer droplet size.

## Nitroxide-mediated Radical Polymerization

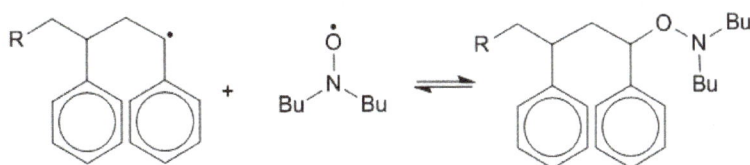

Reversible Termination Reaction          Bu = tert-Butyl

Nitroxide-mediated radical polymerization is a method of radical polymerization that makes use of an alkoxyamine initiator to generate polymers with well controlled stereochemistry and a very low polydispersity index. It is a type of reversible-deactivation radical polymerization.

## Alkoxyamines

The initiating materials for nitroxide-mediated radical polymerization (NMP) are a family of compounds referred to as alkoxyamines. An alkoxyamine can essentially be viewed as an alcohol bound to a secondary amine by an N-O single bond. The utility of this functional group is that under certain conditions, homolysis of the C-O bond can occur, yielding a stable radical in the form of a 2-center 3-electron N-O system and a carbon radical which serves as an initiator for radical polymerization. For the purposes of NMP, the R groups attached to the nitrogen are always bulky, sterically hindering groups and the R group in the O- position forms a stable radical, generally is benzylic for polymerization to occur successfully. NMP allows for excellent control of chain length and structure, as well as a relative lack of true termination that allows polymerization to continue as long as there is available monomer. Because of this it is said to be "living".

## Persistent Radical Effect

The living nature of NMP is due to the persistent radical effect (PRE). The PRE is a phenomenon observable in some radical systems which leads to the highly favored formation of one product to the near exclusion of other radical couplings due to one of the radical species being particularly stable, existing in greater and greater concentrations as the reaction progresses while the other one is transient, reacting quickly with either itself in a termination step or with the persistent radical to form a desired product. As time goes on, a higher concentration of the persistent radical is present, which couples reversibly with itself, meaning that any of the transient radical still present tends to couple with the persistent radical rather than itself due to greater availability. This leads to a greater proportion of cross-coupling than self-coupling in radical species.

In the case of a nitroxide-mediated polymerization reaction, the persistent radical is the nitroxide species and the transient radical is always the carbon radical. This leads to repeated coupling of the nitroxide to the growing end of the polymer chain, which would ordinarily be considered a termination step, but is in this case reversible. Because of the high rate of coupling of the nitroxide to the growing chain end, there is little coupling of two active growing chains, which would be an irreversible terminating step limiting the chain length. The nitroxide binds and unbinds to the growing chain, protecting it from termination steps. This ensures that any available monomer can be easily scavenged by active chains. Because this polymerization process does not naturally self-terminate, this polymerization process is described as "living," as the chains continue to grow under suitable reaction conditions whenever there is reactive monomer to "feed" them. Because of the PRE, it can be assumed that at any given time, almost all of the growing chains are "capped" by a mediating nitroxide, meaning that they dissociate and grow at very similar rates, creating a largely uniform chain length and structure.

## Nitroxide Stability

As stated above, nitroxide radicals are effective mediators of well-controlled radical polymerization because they are quite stable, allowing them to act as persistent radicals in a reaction mixture. This stability is a result of their unique structure. In most diagrams, the radical is depicted on the oxygen, but another resonance structure exists which is more helpful in explaining their stability in which the radical is on the nitrogen, which has a double bond to the oxygen. In addition to this resonancestability, nitroxides used in NMRP always contain bulky, sterically hindering groups in the R1 and R2 positions. The significant steric bulk of these substituents entirely prevents radical coupling in the N-centered resonance form while significantly reducing it in the O-centered form. These bulky groups contribute stability, but only if there is no resonance provided by allyl or aromatic groups α to the N. These result in decreased stability of the nitroxide, presumably because they offer less sterically hindered sites for radical coupling to take place. The resulting inactivity of the radical makes hemolytic cleavage of the alkoxyamine quite fast in more sterically hindered species.

## Nitroxide Choice

The choice of a specific nitroxide species to use has a large effect on the efficacy of an attempted polymerization. An effective polymerization (fast rate of chain growth, consistent chain length) results from a nitroxide with a fast C-O homolysis and relatively few side reactions. A more polar solvent lends itself better to C-O homolysis, so polar solvents which cannot bind to a labile nitroxide are the most effective for NMP. It is generally agreed that the structural factor that has the greatest effect on the ability of a nitroxide to mediate a radical polymerization is steric bulk. Generally speaking, greater steric bulk on the nitroxide leads to greater strain on the alkoxyamine, leading to the most easily broken bond, the C-O single bond, cleaving homolytically.

## Ring Size

In the case of cyclic nitroxides, five-membered ring systems have been shown to cleave more slowly than six-membered rings and acyclic nitroxides with t-butyl moieties as their R groups cleaved fastest of all. This difference in the rate of cleavage was determined to result not from a difference in C-O bond lengths, but in the difference of C-O-N bond angle in the alkoxyamine. The smaller the bond angle the greater the steric interaction between the nitroxide and the alkyl fragment and the more easily the initiator species broke apart.

## Steric Bulk

The efficiency of polymerization increases more and more with increased steric bulk of the nitroxide up to a point. TEMPO ((2,2,6,6-Tetramethylpiperidin-1-yl)oxyl) is capable of inducing the polymerization of styrene and styrene derivatives fairly easily, but is not sufficiently labile to induce polymerization of butyl acrylate under most conditions. TEMPO derivatives with even bulkier groups at the positions α to N have a rate of homolysis great enough to induce NMP of butyl acrylate, and the bulkier the α groups, the faster polymerization occurs. This indicates that the steric bulk of the nitroxide fragment can be a good indicator of the strength of an alkoxyamine initiator, at least up to a point. The equilibrium of its homolysis and reformation favors the radical form to the extent that recombination to reform an alkoxyamine over the course of NMP occurs too slowly to maintain control of chain length.

## Preparation Methods

Because TEMPO, which is commercially available, is a sufficient nitroxide mediator for the synthesis of polystyrene derivatives, the preparation of alkoxyamine initiators for NMP of copolymers is in many cases a matter of attaching a nitroxide group (TEMPO) to a specifically synthesized alkyl fragment. Several methods have been reported to achieve this transformation.

## Jacobsen's Catalyst

Jacobsen's catalyst is a manganese-based catalyst commonly used for the stereoselective epoxidation of alkenes. This epoxidation proceeds by a radical addition mechanism, which can be taken advantage of by introducing the radical TEMPO group into the reaction mixture. After treatment with a mild reducing agent such as sodium borohydride, this yields the product of a Markovnikov addition of nitroxide to the alkene. Jacobsen's catalyst is fairly mild, and a wide variety of functionalities on the alkene substrate can be tolerated. Practical yields are not necessarily as high as those reported by Dao et al., however.

## Hydrazine

An alternative method is to react a substrate with a C-Br bond at the desired location of the nitroxide with hydrazine, generating an alkyl substituted hydrazine which is then exposed to a nitroxide radical and a mild oxidating agent such as lead dioxide. This generates a carbon-centered radical which couples with the nitroxide to generate the desired alkoxyamine. This method has the disadvantage of being relatively inefficient for some species, as well as the inherent danger of having to work with extremely toxic hydrazine and the inconvenience of having to run reactions in inert atmosphere.

## Treatment of Aldehydes with Hydrogen Peroxide

Yet another published alkoxyamine synthesis involves treatment of aldehydes with hydrogen peroxide, which adds to the carbonyl group. The resulting species rearranges in situ in the presence of CuCl forming formic acid and the desired alkyl radical, which couples with tempo to produce the target alkoxyamine. The reaction appears to give fairly good yields and tolerates a variety of functional groups in the alkyl chain.

## Electrophilic Bromination and Nucleophilic Attack

A synthesis has been described by Moon and Kang consisting of a one-electron reduction of a nitroxide radical in metallic sodium to yield a nucleophilic nitroxide. The nitroxide nucleophile is then added to an appropriate alkyl bromide, yielding the alkoxyamine by a simple SN2 reaction. This technique has the advantage of requiring only the appropriate alkyl bromide to be synthesized without requiring inconvenient reaction conditions and extremely hazardous reagents like Braso et al.'s method.

## Suspension Polymerization

Suspension polymerization is a heterogeneous radical polymerization process.

Suspension polymerization is used for the commercial manufacture of many important polymers including poly(vinyl chloride), poly(methyl methacrylate), expandable polystyrene, styrene–acrylonitrile copolymers and a variety of ion exchange resins. In suspension polymerization, drops of a monomer-containing phase are dispersed in a continuous liquid phase and polymer is produced inside the drops. In many cases, the monomer contains no diluent and the chemical reactions that occur inside the drops are very similar to those that are found in bulk polymerization. In most suspensions, polymer is formed via a chain reaction mechanism that includes the following steps:

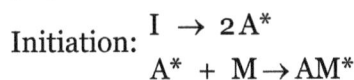

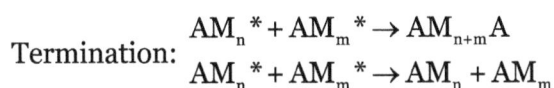

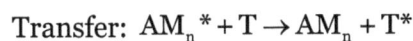

Initiation:
$$I \rightarrow 2A^*$$
$$A^* + M \rightarrow AM^*$$

Propagation: $AM_n^* + M \rightarrow AM_{n+1}^*$

Termination:
$$AM_n^* + AM_m^* \rightarrow AM_{n+m}A$$
$$AM_n^* + AM_m^* \rightarrow AM_n + AM_m$$

Transfer: $AM_n^* + T \rightarrow AM_n + T^*$

Here, M is the monomer and $A^*$ could be an anion, a cation or a free radical. In most industrial processes M is a vinyl compound and a free-radical chain mechanism is used. Then, the growing polymer chains, $AM_n^*$, are written as $AM_n \cdot$. That species has a short life-time (usually $\ll$ 1 sec) and completed polymer molecules are formed throughout the process. The generation of radicals, $A\cdot$, is usually induced by thermal decomposition of an organic initiator, I, that is soluble in the monomer. Organic peroxides are often used as initiators. T represents any species that reacts as a chain transfer agent. T can be monomer, polymer, a solvent or a species that is added specifically to function as a chain transfer agent. If $T^*$ is sufficiently active it can behave as $A^*$ and initiate a new polymer chain. In some cases, such as the polymerization of vinyl chloride, chain transfer to monomer is significant and it has a major effect on the average molecular weight of the polymer.

In most industrial suspension polymerization agitated batch (or semi-batch) reactors are used and the continuous phase is aqueous. That is advantageous because the process is often exothermic and good heat transfer from the reactor is required. The ratio of surface area to volume is relatively high for small drops so that the rate of heat transfer to the aqueous phase is high. Although drop viscosity may increase substantially, the overall viscosity of the suspension is usually much lower than that which is encountered in the equivalent bulk polymerization. Consequently, agitation of the reactor contents is possible and heat transfer via the aqueous phase to the reactor wall is good. Also, high conversions of monomer to polymer can be achieved inside the drops whereas, in bulk polymerization, increasing viscosity of the polymer-monomer solution often limits the extent of monomer conversion.

Bulk copolymerization may become difficult to control if cross-linking or copolymer precipitation occurs; then, a suspension process may then be the only feasible way in which the copolymerization can be carried out.

Suspension polymerization is particularly useful when the final polymer is required to be in the form of small "beads" (which often have the same size distribution as the drops from which they are formed). However, product contamination can be a problem if the drop stabilizers cannot be removed. Suspension polymerization usually requires larger reactor volumes than bulk processes because the vessels are usually half full with water.

The attainment of high monomer conversion can affect the reaction kinetics. From the reaction scheme shown above, it can be shown that the rate of homogeneous polymerization is given by the expression:

$$R_p = k_p C_M \left( \frac{2 f k_d C_1}{k_t} \right)^{1/2}$$

Where $CI$ is the concentration of the initiator and $CM$ is the monomer concentration. Here, $kp$ is the propagation rate coefficient, $kd$ is the initiator decomposition rate coefficient and $f$ is an efficiency factor. In equation, the overall chain termination rate coefficient, $kt$, is derived from the rate coefficients of the two chain termination steps that are shown in the above reaction scheme. At high polymer concentrations, chain termination is often diffusion-controlled and the value of $kt$ diminishes substantially. Radical diffusion can depend on solution viscosity, polymer volume fraction and polymer molecular weight. The latter three entities are interrelated in complicated ways but the effects of viscosity on polymerization rate can be distinguished from the effects of polymer volume fraction. The value of $f$ may depend on polymer content and the value of $kp$ may also decrease. From equation, it can be seen that the reduction in $kt$ leads to an increase in the polymerization rate, a phenomenon often described as a "gel effect".

## Suspending Agents

In the absence of a drop stabilizer, the suspension would be unstable and the monomer/polymer drops would coalesce and become large. That is undesirable because, often, it is necessary to obtain a specific size distribution for the final polymer particles. Therefore, control of drop size, and of drop stability, during polymerization becomes important. Drop stability depends largely on the nature of the drop stabilizer (or suspending agent). Adsorption of stabilizer molecules on the outer drop surface can reduce the interfacial tension and, hence, lower the energy required for drop formation. However, drop stability against coalescence depends largely on the ability of the stabilizer to form a protective film at the interface. Increasing the stabilizer concentration continues to improve the elastic properties of the drops until a "critical surface coverage" is attained; further increases then have a very little effect on the drop stability.

Water-miscible polymers, both naturally-occurring and synthetic, are often used as drop stabilizers. Initially, these materials are dispersed in the continuous phase; subsequent migration to the surfaces of newly-created monomer drops may be rapid but the development of drop stability may be slow because rearrangement of stabilizer molecules on the drop surface is necessary . When partially hydrolyzed polyvinyl acetate (PVA) is used as a stabilizer its behavior depends on the extent to which the acetate groups are hydrolyzed. Good drop stabilization can be achieved in aqueous media when the degree of hydrolysis (DH) is between 70% and 80 %; then, drops can retain their integrity even when agitation levels are reduced. PVAs with a DH less than 60% are poor drop stabilizers in aqueous media but they can affect polymer morphology inside the (non-aqueous) drops. That is important in the suspension polymerization of vinyl chloride (VC). In that case, small particles of poly(vinyl chloride) (PVC) precipitate inside the monomer drops because PCV and VC are almost immiscible. Therefore, a mixture of two stabilizers is often used; a "primary stabilizer" which protects the drops from coalescence and a "secondary stabilizer" which effects the behavior of the PVC particles inside the drops and increases polymer porosity. The addition of a secondary stabilizer can also effect the particle size distribution of the polymer particles. PVC porosity can also be increased by using non-ionic surfactants as secondary stabilizers. PVA can become grafted onto polymer that is formed inside the drops so that a "skin" forms on the final particle surface . Formation of that skin, which is difficult to remove, can affect the final polymer properties. With some monomers, product contamination can be avoided by using alternative suspending agents such as salts of polymethacrylic acid which are not grafted on the particle surface and can be removed from the final polymer product with an aqueous wash. If the initiator in suspension polymerization is slightly soluble in water then simultaneous emulsion polymerization may occur when free stabilizer remains in the continuous phase.

## Suspension co-polymerization

Functional groups can be introduced via co-polymerization with appropriate monomers but control and prediction of co-polymer compositions in suspension polymerization can be difficult if one, or more, of the monomers is partially soluble in the continuous phase. Then, the actual monomer concentrations in the drops may be unknown so that idealized relationships for predicting co-polymer compositions, which apply to homogeneous systems, are of little use unless appropriate partition coefficients for the two phases are available. Apparent reactivity ratios, obtained directly from suspension polymerization experiments will be different from those expected for the equivalent bulk processes if some monomer migrates to the continuous phase. In some cases, when the continuous phase is aqueous, models that allow for water solubility of monomers have been developed.

## Drop Formation and Stability

Control of drop size distribution in suspension polymerisation can be important. In many cases, the average drop diameters (and final average particles sizes) lie between 10 and 100 μm but larger diameters might be produced if the polymer particles are to be used directly as beads.

The physical conditions in a suspension polymerization reactor affect the drop size distribution significantly. Drop breakage in agitated suspensions can be caused either by frictional forces (via viscous shear) or by inertial forces (via turbulence) . In industrial suspension polymerisation, the volume fraction of dispersed phase is usually high and drop break-up is accompanied by drop coalescence. Thus, the average drop size and the drop size distribution are both influenced by drop breakage and by drop coalescence.

## Suspensions with Turbulent Flow

In large vessels, agitated aqueous suspensions are often turbulent. If turbulence is isotropic and the diameter exceeds the Kolmogorov length then turbulent pressure fluctuations will cause drop breakage. By applying some simplifying assumptions, the average drop size is sometimes given by equation.

$$\frac{d_{32}}{D} = a\left(1 + b\phi\right)We^{-0.6}$$

Here, $d_{32}$ is the Sauter–mean drop diameter, $\varphi$ is the volume fraction of the dispersed phase, $D$ is the impeller diameter and $a$ and $b$ are constants . The Weber number, We, is given by

$$We = \frac{\rho_m N^2 D^3}{\sigma}$$

Where $N$ is the stirrer speed, $\rho_m$ is the dispersion density and $\sigma$ is the interfacial tension.

Equation reflects a balance between inertial forces and interfacial forces but, even when drops are formed via turbulence, application of equation is limited to suspensions in which the dispersed phase has a low viscosity and the drop concentration is low (so that drop coalescence is not significant). Wang and Calabrese. showed that, even when turbulence is important, drop break-up can be opposed by both interfacial forces and viscous forces and that the influence of interfacial tension on drop breakage decreases as the dispersed-phase viscosity increases. Drop sizes can take some time to be established and, if polymerisation occurs during that time, the drop viscosity may increase, consequently rates of drop break-up and coalescence will be reduced. Therefore, drop breakage can be a complex process and the polymer particles can have a broad size distribution. In the suspension polymerisation of methyl methacrylate drop viscosity increases significantly and four separate stages have been identified in the drop formation process. In the suspension polymerization of styrene, Konno et al. found that drop coalescence was important and that the Sauter mean diameter increased as the polymer viscosity increased. They also concluded that the stabilizer does not effectively prevent the coalescence of drops with diameters that are larger than those predicted from Weber Number correlations.

The drop coalescence rate can be related to the drop collision frequency and to the coalescence efficiency. Coalescence may occur if drops adhere for sufficient time to allow them to deform, and to permit drainage of the continuous phase that is trapped between them . By taking account of these events, expressions can be obtained for the coalescence efficiency.

## Copolymerization

A homopolymer is defined as a polymer which is obtained by linking together only similar types of molecules (or monomers), whereas a copolymer is obtained by linking two or more different types of monomers in the same chain. There are many techniques of polymerization such as condensation, polymerization and copolymerization.

When a mixture of more than one (or different) monomeric species is allowed to polymerize and form a copolymer , then we call this process as copolymerization. Example- Nylon 66 is a copolymer of hexamethylenediamine and adipic acid. There are four different types of copolymers existing today:

- Alternating copolymers: The copolymers which have two units placed at alternating positions. For example, we place a unit X then Y, then X and then again Y and the chain goes on.

—— A —— B —— A —— B ——

The above figure shows the arrangement of alternating copolymers.

- Periodic copolymers: The units of both X and Y is arranged in a repetitive sequence. The arrangement is as shown below:

$$\left( A - B - A - B - B - A - A - A \right)_n$$

- Statistical copolymers: The arrangement of polymers in this case follows a statistical rule. When the probability of finding a given monomer at a particular point is equal to the mole fraction of that monomer in the same chain, then the polymer is referred to as a random copolymer. It has the following arrangement:

$$- A - B - B - B - A - B - A - B -$$

- Block copolymers: When two or more polymers are linked by homopolymers then they are called as block copolymers.

$$- B - B - B - B - A - A - A - A -$$

The above arrangement shows the structure of the block copolymers.

## The Copolymer Equation

It can be shown that the rate of change of monomer concentration in any copolymerization is given by the equation:

$$\frac{d[M_1]}{d[M_2]} = \frac{[M_1]}{[M_2]} \cdot \frac{r_1[M_1] + [M_2]}{r_2[M_2] + [M_1]}$$

where $[M_1]$ and $[M_2]$ are the concentrations of monomers 1 and 2 at any instant and $r_1$ and $r_2$, are reactivity ratios. The reactivity ratios represent the rate at which one type of growing chain end adds on to a monomer of the same structure relative to the rate at which it adds on to the alternative monomer. The copolymer equation can be used to predict chain structure in the three different ways, already mentioned.

The formation of regular alternating copolymers of the type ABAB is favoured when each growing radical prefers to add to monomer of the opposite type. In this case

$$r_1 \approx r_2 \approx 0$$

and Equation $\dfrac{d[M_1]}{d[M_2]} = \dfrac{[M_1]}{[M_2]} \cdot \dfrac{r_1[M_1] + [M_2]}{r_2[M_2] + [M_1]}$ therefore becomes

$$\frac{d[M_1]}{d[M_2]} = 1 \text{ or } d[M_1] = d[M_2]$$

In other words both monomers will disappear from the reaction vessel at the same rate.

An ideal copolymer will tend to form when each type of chain end shows an equal preference for adding on to either monomer. In this case,

$$r_1 = \frac{1}{r_2}$$

and the copolymer equation becomes

$$\frac{d[M_1]}{d[M_2]} = r_1 \cdot \frac{[M_1]}{[M_2]}$$

Hence composition depends on the relative amounts of monomer present at any time and the relative reactivities of the two monomers.

Finally, block copolymers are formed when the growing chain end has a marked preference for adding on to the same kind of monomer. In this case

$$r_1 > 1$$
$$r_2 > 2$$

As can be seen from Table, this is rarely achieved in free radical copolymerization. However, it is possible to form block structures in anionic polymerization simply by feeding different monomers to the living polymer. Step growth copolymerizations produce ideal (random) copolymers since in this special case $r_1 = r_2 = 1$.

Table, Reactivity ratios for free radical chain growth polymerization:

| Monomer 1 | Monomer 2 | $r_1$ | $r_2$ |
|---|---|---|---|
| acrylonitrile | 1,3-butadiene | 0.02 | 0.3 |
| | methyl methacrylate | 0.15 | 1.22 |
| | styrene | 0.04 | 0.40 |
| | vinyl acetate | 4.2 | 0.05 |
| | vinyl chloride | 2.7 | 0.04 |
| 1,3-butadiene | methyl methacrylate | 0.75 | 0.25 |
| | styrene | 1.35 | 0.78 |
| | vinyl chloride | 8.8 | 0.035 |
| methyl methacrylate | styrene | 0.46 | 0.52 |
| | vinyl acetate | 20 | 0.015 |
| | vinyl chloride | 10 | 0.1 |
| styrene | vinyl acetate | 55 | 0.01 |
| | vinyl chloride | 17 | 0.02 |
| vinyl acetate | vinyl chloride | 0.23 | 1.68 |

## Commercial Copolymers

The main reason for copolymerizing different monomers is to adjust the physical properties of a given homopolymer to meet a specific demand. SBR elastomer, for example Table, based on 24 wt% styrene monomer shows better mechanical properties and better resistance to degradation than polybutadiene alone. By increasing the styrene content to 35 per cent, a high hysteresis (energy absorbing) material ideal for tyre treads is produced. Another example is nitrile rubber, which is produced by a free radical emulsion copolymerization of butadiene and acrylonitrile to make an oil-resistant rubber suitable for oil and petrol lines.

A second reason for copolymerization is to enhance the chemical reactivity of a polymer, particularly to aid crosslinking. Conventional vulcanization in rubbers is brought about by forming sulphur crosslinks at or near double bonds in the chain. In polyisobutylene where the main chain repeat unit is there are no such bonds. So the isobutylene monomer is copolymerized with a few weight percent isoprene units to make IIR (butyl rubber) which can be vulcanized easily.

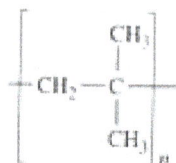

$$\left[ CH_2 - \underset{\underset{CH_3}{|}}{\overset{\overset{CH_3}{|}}{C}} \right]_n$$

This is also the reason why EPDM rubber consists of no less than three different monomer units copolymerized together (ethylene, propylene and a diene) using Ziegler-Natta catalysts. The copolymer structure is random, so crystallinity is low and the material behaves like a rubber when vulcanized across the diene double bonds.

To show the dramatic effect of copolymer structure on physical properties, consider the change from random SBR copolymer to a block copolymer of exactly the same chemical composition but where the styrene and butadiene parts are effectively homopolymer chains linked at two points:

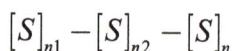

$$[S]_{n1} - [S]_{n2} - [S]_{n}$$

The material behaves like a vulcanized butadiene rubber without the need for chemical crosslinking since the styrene chains segregate together to form small islands or domains within the structure. Such so-called thermoplastic elastomers (TPEs) today form an important growth area for new polymers because of the process savings in manufacture that can be achieved with their use.

Among rigid thermoplastics, the most widely used copolymers are those of styrene and they include ABS, HIPS and SAN. Both HIPS and ABS are graft block copolymers where the elastomeric side chains are deliberately introduced to improve the toughness of the material.

# Step-growth Polymerization

A step-growth polymerization is a stepwise reaction between bi-functional or multi-functional monomers in which a high-molecular-weight polymer is formed after a large number of steps. Many naturally and synthetic polymers are produced by step-growth polymerization including polyesters, polyethers, urethanes, epoxies, and polyamides (see table below). Two well-known examples are the reaction of dicarboxylic acids with diamines to form polyamides (Nylon) and the reaction of organic diacids with alcohols to form polyesters, like polyethylene terephthalate (PET). Due to the nature of the polymerization mechanism, the reaction has to proceed for a long time to achieve high molecular weight polymers. The easiest way to visualize the step-growth mechanism is a crowd of people reaching out to hold their hands to form human chains — each person has two hands (= two reactive sites).

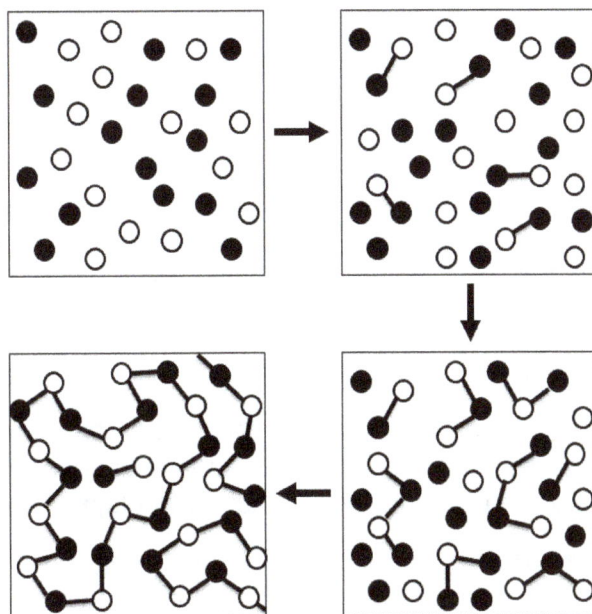

If the monomers have more than two reactive sites branched polymers are produced. Some important step-growth polymerizations are shown below.

| EXAMPLES OF STEP-GROWTH POLYMERIZATIONS | | |
|---|---|---|
| Functional Groups | Linkage | Polymer Type |
| -OH + -COOH | -C(=O)-O- | Polyester |
| -OH + -NCO | -O-C(=O)-NH- | Polyurethane |
| -NH$_2$ + -NCO | -NH-C(=O)-NH- | Polyurea |
| -NH$_2$ + -COOH | -NH-C(=O)- | Polyamide |
| -OH + -OH | -O- | Polyether |

The two most important step-growth polymerizations are *condensation* and *addition* polymerization. In the case of a condensation reaction, two monomers combine with the loss of a small molecule, usually an alcohol, a water molecule or an acid, whereas an addition reaction involves only the rearrangement of the electrons of a double bond to form a single bond with another molecule.

## Classes of Step-growth Polymers

Classes of step-growth polymers are:

- Polyester has high glass transition temperature $T_g$ and high melting point $T_m$, good mechanical properties to about 175 °C, good resistance to solvent and chemicals. It can exist as fibers and films. The former is used in garments, felts, tire cords, etc. The latter appears in magnetic recording tape and high grade films.

- Polyamide (nylon) has good balance of properties: high strength, good elasticity and abrasion resistance, good toughness, favorable solvent resistance. The applications of polyamide include: rope, belting, fiber cloths, thread, substitute for metal in bearings, jackets on electrical wire.

- Polyurethane can exist as elastomers with good abrasion resistance, hardness, good resistance to grease and good elasticity, as fibers with excellent rebound, as coatings with good resistance to solvent attack and abrasion and as foams with good strength, good rebound and high impact strength.

- Polyurea shows high $T_g$, fair resistance to greases, oils, and solvents. It can be used in truck bed liners, bridge coating, caulk and decorative designs.

- Polysiloxane are available in a wide range of physical states—from liquids to greases, waxes, resins, and rubbers. Uses of this material are as antifoam and release agents, gaskets, seals, cable and wire insulation, hot liquids and gas conduits, etc.

- Polycarbonates are transparent, self-extinguishing materials. They possess properties like crystalline thermoplasticity, high impact strength, good thermal and oxidative stability. They can be used in machinery, auto-industry, and medical applications. For example, the cockpit canopy of F-22 Raptor is made of high optical quality polycarbonate.

- Polysulfides have outstanding oil and solvent resistance, good gas impermeability, good resistance to aging and ozone. However, it smells bad, and it shows low tensile strength as well as poor heat resistance. It can be used in gasoline hoses, gaskets and places that require solvent resistance and gas resistance.

- Polyether shows good thermoplastic behavior, water solubility, generally good mechanical properties, moderate strength and stiffness. It is applied in sizing for cotton and synthetic fibers, stabilizers for adhesives, binders, and film formers in pharmaceuticals.

- Phenol formaldehyde resin (Bakelite) have good heat resistance, dimensional stability as well as good resistance to most solvents. It also shows good dielectric properties. This material is typically used in molding applications, electrical, radio, televisions and automotive parts where their good dielectric properties are of use. Some other uses include: impregnating paper, varnishes, decorative laminates for wall coverings.

- Poly-Triazole polymers are produced from monomers which bear both an alkyne and azide functional group. The monomer units are linked to each other by the a 1,2,3-triazole group; which is produced by the 1,3-Dipolar cycloaddition, also called the Azide-alkyne Huisgen cycloaddition. These polymers can take on the form of a strong resin, or a gel. With oligopeptide monomers containing a terminal alkyne and terminal azide the resulting clicked peptide polymer will be biodegradable due to action of endopeptidases on the oligopeptide unit.

## Branched Polymers

A monomer with functionality of 3 or more will introduce branching in a polymer and will ultimately form a cross-linked macrostructure or network even at low fractional conversion. The point at which a tree-like topology transits to a network is known as the gel point because it is signalled by an abrupt change in viscosity. One of the earliest so-called thermosets is known as bakelite. It is not always water that is released in step-growth polymerization: in acyclic diene metathesis or ADMET dienes polymerize with loss of ethylene.

## Kinetics

The kinetics and rates of step-growth polymerization can be described using a polyesterification mechanism. The simple esterification is an acid-catalyzed process in which protonation of the acid is followed by interaction with the alcohol to produce an ester and water. However, there are a few assumptions needed with this kinetic model. The first assumption is water (or any other condensation product) is efficiently removed. Secondly, the functional group reactivity's are independent of chain length. Finally, it is assumed that each step only involves one alcohol and one acid.

$$\frac{1}{1-p^{n-1}} = 1 + (n-1)kt[COOH]^{n-1}$$

This is a general rate law degree of polymerization for polyesterification where n= reaction order.

## Self-catalyzed Polyesterification

If no acid catalyst is added, the reaction will still proceed because the acid can act as its own catalyst. The rate of condensation at any time t can then be derived from the rate of disappearance of -COOH groups and

$$rate = \frac{-d[\text{COOH}]}{dt} = k[\text{COOH}]^2[\text{OH}]$$

The second-order [\ce{COOH}] term arises from its use as a catalyst, and k is the rate constant. For a system with equivalent quantities of acid and glycol, the functional group concentration can be written simply as

$$rate = \frac{-d[\text{COOH}]}{dt} = k[\text{COOH}]^3$$

After integration and substitution from Carothers equation, the final form is the following

$$\frac{1}{(1-p)^2} = 2kt[\text{COOH}]^2 + 1 = X_n^2$$

For a self-catalyzed system, the number average degree of polymerization (Xn) grows proportionally with $\sqrt{t}$.

## External Catalyzed Polyesterification

The uncatalyzed reaction is rather slow, and a high $X_n$ is not readily attained. In the presence of a catalyst, there is an acceleration of the rate, and the kinetic expression is altered to

$$\frac{-d[\text{COOH}]}{dt} = k[\text{COOH}][\text{OH}]$$

Which is kinetically first order in each functional group. Hence,

$$\frac{-d[\text{COOH}]}{dt} = k[\text{COOH}]^2$$

and integration gives finally

$$\frac{1}{1-p} = 1 + [\text{COOH}]kt = X_n$$

For an externally catalyzed system, the number average degree of polymerization grows proportionally with $t$.

## Molecular Weight Distribution in Linear Polymerization

The product of a polymerization is a mixture of polymer molecules of different molecular weights. For theoretical and practical reasons it is of interest to discuss the distribution of molecular weights in a polymerization. The molecular weight distribution (MWD) had been derived by Flory by a statistical approach based on the concept of equal reactivity of functional groups.

## Probability

Step-growth polymerization is a random process so we can use statistics to calculate the probability of finding a chain with x-structural units ("x-mer") as a function of time or conversion.

$$x\,AA + x\,BB \rightarrow AA\text{-}(BB\text{-}AA)_{x-1}\text{-}BB$$

$$x\,AB \rightarrow A\text{-}(B\text{-}A)_{x-1}\text{-}B$$

Probability that an 'A' functional group has reacted

$$p^{x-1}$$

Probability of finding an 'A' unreacted

$$(1-p)$$

Combining the above two equations leads to:

$$P_x = (1-p)p^{x-1}$$

Where $P_x$ is the probability of finding a chain that is x-units long and has an unreacted 'A'. As x increases the probability decreases.

## Number Fraction Distribution

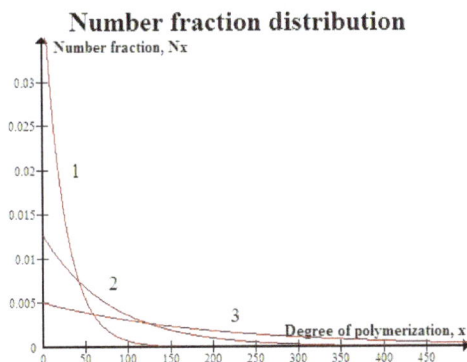

Number-fraction distribution curve for linear polymerization. Plot 1, p=0.9600; plot 2, p=0.9875; plot 3, p=0.9950.

The number fraction distribution is the fraction of x-mers in any system and equals the probability of finding it in solution.

$$\frac{N_x}{N} = (1-p)p^{x-1}$$

Where N is the total number of polymer molecules present in the reaction.

## Weight Fraction Distribution

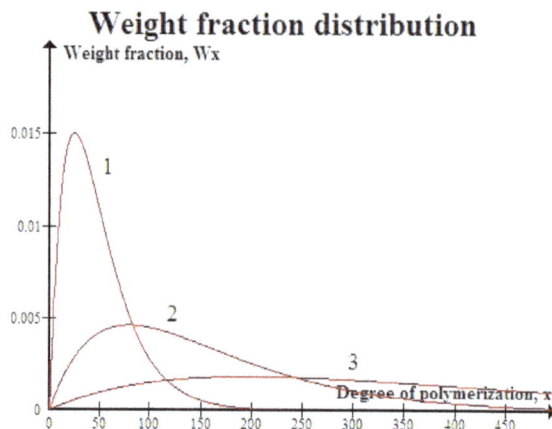

Weight fraction distribution plot for linear polymerization. Plot 1, p=0.9600; plot 2, p=0.9875; plot 3, p=0.9950.

The weight fraction distribution is the fraction of x-mers in a system and the probability of finding them in terms of mass fraction.

$$\frac{W_x}{W_o} = \frac{xN_xM_o}{N_oM_o} = \frac{xN_x}{N_o} = x\frac{N_x}{N}\frac{N}{N_o}$$

Notes:

- $M_o$ is the molar mass of the repeat unit,

- $N_o$ is the initial number of monomer molecules,

- and $N$ is the number of unreacted functional groups

Substituting from the Carothers equation

$$X_n = \frac{1}{1-p} = \frac{N_o}{N}$$

We can now obtain:

$$\frac{W_x}{W_o} = x(1-p)^2 p^{x-1}$$

## PDI

The polydispersity index (PDI), is a measure of the distribution of molecular mass in a given polymer sample.

$$PDI = \frac{M_w}{M_n}$$

However, for step-growth polymerization the Carothers equation can be used to substitute and rearrange this formula into the following.

$$PDI = 1 + p$$

Therefore, in step-growth when p=1, then the PDI=2.

## Molecular Weight Control in Linear Polymerization

## Need for Stoichiometric Control

There are two important aspects with regard to the control of molecular weight in polymerization. In the synthesis of polymers, one is usually interested in obtaining a product of very specific molecular weight, since the properties of the polymer will usually be highly dependent on molecular weight. Molecular weights higher or lower than the desired weight are equally undesirable. Since the degree of polymerization is a function of reaction time, the desired molecular weight can be obtained by quenching the reaction at the appropriate time. However, the polymer obtained in this manner is unstable in that it leads to changes in molecular weight because the ends of the polymer molecule contain functional groups that can react further with each other.

This situation is avoided by adjusting the concentrations of the two monomers so that they are slightly nonstoichiometric. One of the reactants is present in slight excess. The polymerization then proceeds to a point at which one reactant is completely used up and all the chain ends possess the same functional group of the group that is in excess. Further polymerization is not possible, and the polymer is stable to subsequent molecular weight changes.

Another method of achieving the desired molecular weight is by addition of a small amount of monofunctional monomer, a monomer with only one functional group. The monofunctional monomer, often referred to as a chain stopper, controls and limits the polymerization of bifunctional monomers because the growing polymer yields chain ends devoid of functional groups and therefore incapable of further reaction.

## Quantitative Aspects

To properly control the polymer molecular weight, the stoichiometric imbalance of the bifunctional monomer or the monofunctional monomer must be precisely adjusted. If the nonstoichiometric imbalance is too large, the polymer molecular weight will be too low. It is important to understand the quantitative effect of the stoichiometric imbalance of reactants on the molecular weight. Also, this is necessary in order to know the quantitative effect of any reactive impurities that may be present in the reaction mixture either initially or that are formed by undesirable side reactions. Impurities with A or B functional groups may drastically lower the polymer molecular weight unless their presence is quantitatively taken into account.

More usefully, a precisely controlled stoichiometric imbalance of the reactants in the mixture can provide the desired result. For example, an excess of diamine over an acid chloride would eventually produce a polyamide with two amine end groups incapable of further growth when the acid chloride was totally consumed. This can be expressed in an extension of the Carothers equation as,

$$X_n = \frac{(1+r)}{(1+r-2rp)}$$

where r is the ratio of the number of molecules of the reactants.

$$r = \frac{N_{AA}}{N_{BB}} \text{ were } N_{BB} \text{ is the molecule in excess.}$$

The equation above can also be used for a monofunctional additive which is the following,

$$r = \frac{N_{AA}}{(N_{BB} + 2N_B)}$$

where $N_B$ is the number of monofunction molecules added. The coefficient of 2 in front of $N_B$ is require since one B molecule has the same quantitative effect as one excess B-B molecule.

## Multi-chain Polymerization

A monomer with functionality 3 has 3 functional groups which participate in the polymerization. This will introduce branching in a polymer and may ultimately form a cross-linked macrostructure. The point at which this three-dimensional 3D network is formed is known as the gel point, signaled by an abrupt change in viscosity.

A more general functionality factor $f_{av}$ is defined for multi-chain polymerization, as the average number of functional groups present per monomer unit. For a system containing $N_o$ molecules initially and equivalent numbers of two function groups A and B, the total number of functional groups is $N_o f_{av}$.

$$f_{av} = \frac{\sum N_i \cdot f_i}{\sum N_i}$$

And the modified Carothers equation is

$$x_n = \frac{2}{2 - p f_{av}}, \text{ where p equals to } \frac{2(N_o - N)}{N_o \cdot f_{av}}$$

## Advances in Step-growth Polymers

The driving force in designing new polymers is the prospect of replacing other materials of construction, especially metals, by using lightweight and heat-resistant polymers.

The advantages of lightweight polymers include: high strength, solvent and chemical resistance, contributing to a variety of potential uses, such as electrical and engine parts on automotive and aircraft components, coatings on cookware, coating and circuit boards for electronic and microelectronic devices, etc. Polymer chains based on aromatic rings are desirable due to high bond strengths and rigid polymer chains. High molecular weight and crosslinking are desirable for the same reason. Strong dipole-dipole, hydrogen bond interactions and crystallinity also improve heat resistance. To obtain desired mechanical strength, sufficiently high molecular weights are necessary, however, decreased solubility is a problem. One approach to solve this problem is to introduce of some flexibilizing linkages, such as isopropylidene, C=O, and SO$_2$ into the rigid polymer chain by using an appropriate monomer or comonomer. Another approach involves the synthesis of reactive telechelic oligomers containing functional end groups capable of reacting with each other, polymerization of the oligomer gives higher molecular weight, referred to as chain extension.

## Aromatic Polyether

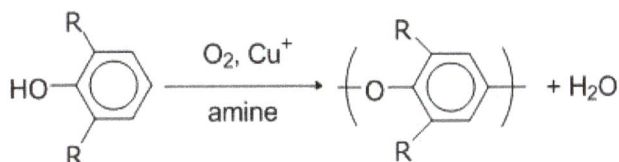

The oxidative coupling polymerization of many 2,6-disubstituted phenols using a catalytic complex of a cuprous salt and amine form aromatic polyethers, commercially referred to as poly(p-phenylene oxide) or PPO. Neat PPO has little commercial uses due to its high melt viscosity. Its available products are blends of PPO with high-impact polystyrene (HIPS).

## Polyethersulfone

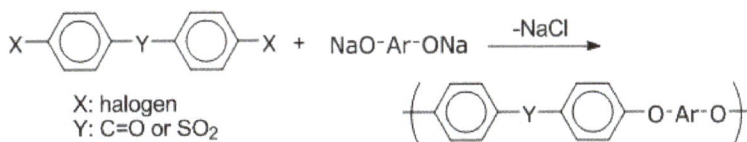

X: halogen
Y: C=O or SO$_2$

Polyethersulfone (PES) is also referred to as polyetherketone, polysulfone. It is synthesized by nucleophilic aromatic substitution between aromatic dihalides and bisphenolate salts. Polyethersulfones are partially crystalline, highly resistant to a wide range of aqueous and organic environment. They are rated for continuous service at temperatures of 240-280 °C. The polyketones are finding applications in areas like automotive, aerospace, electrical-electronic cable insulation.

## Aromatic Polysulfides

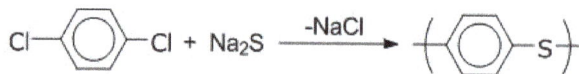

Poly(p-phenylene sulfide) (PPS) is synthesized by the reaction of sodium sulfide with p-dichlorobenzene in a polar solvent such as 1-methyl-2-pyrrolidinone (NMP). It is inherently flame-resistant and stable toward organic and aqueous conditions; however, it is somewhat susceptible to oxidants. Applications of PPS include automotive, microwave oven component, coating for cookware when blend with fluorocarbon polymers and protective coatings for valves, pipes, electromotive cells, etc.

## Aromatic Polyimide

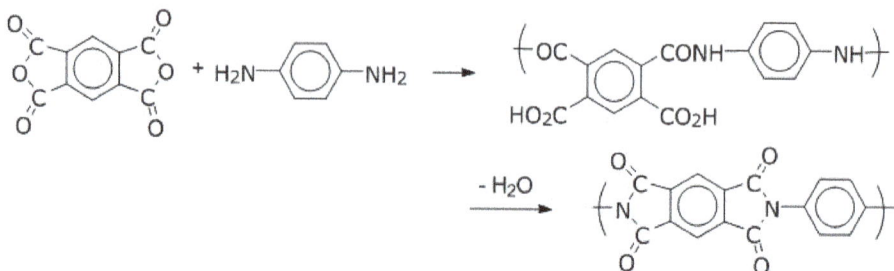

Aromatic polyimides are synthesized by the reaction of dianhydrides with diamines, for example, pyromellitic anhydride with p-phenylenediamine. It can also be accomplished using diisocyanates in place of diamines. Solubility considerations sometimes suggest use of the half acid-half ester of the dianhydride, instead of the dianhydride itself. Polymerization is accomplished by a two-stage process due to the insolubility of polyimides. The first stage forms a soluble and fusible high-molecular-weight poly(amic acid) in a polar aprotic solvent such as NMP or N,N-dimethylacetamide. The poly(amic aicd) can then be processed into the desired physical form of the final plymer product (e.g., film, fiber, laminate, coating) which is insoluble and infusible.

## Telechelic Oligomer Approach

Telechelic oligomer approach applies the usual polymerization manner except that one includes a monofunctional reactant to stop reaction at the oligomer stage, generally in the 50-3000 molecular weight. The monofunctional reactant not only limits polymerization but end-caps the oligomer with functional groups capable of subsequent reaction to achieve curing of the oligomer. Functional groups like alkyne, norbornene, maleimide, nitrite, and cyanate have been used for this purpose. Maleimide and norbornene end-capped oligomers can be cured by heating. Alkyne, nitrile, and cyanate end-capped oligomers can undergo cyclotrimerization yielding aromatic structures.

# Chain-growth Polymerization

Chain growth polymerization is basically a three-stage process, involving initiation of active molecules, their propagation and termination of the active chain ends.

## Initiation

Initiation is the mechanism which starts the polymerization process. Vinyl monomers are quite easily polymerized by a variety of activating methods. Styrene, for example, can be converted to solid polymer simply by heating, and ultraviolet light can have exactly the same effect. Usually, however, an activating agent is used. This is an unstable chemical which produces active species that attack the monomer. A good example is benzoyl peroxide which splits up when heated:

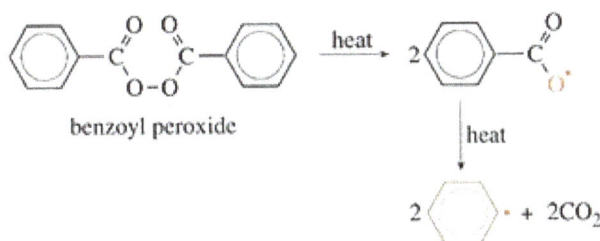

benzoyl peroxide

The formulae of the products are written with a dot alongside to show that they are free radicals. A free radical is a molecule in which there is an unpaired electron. This free radical is very reactive and will attack monomer molecules when introduced into a polymerization vessel. Thus, as benzoyl peroxide is added to styrene (a reaction used with GRP), the peroxide splits to make free radicals, which react as follows:

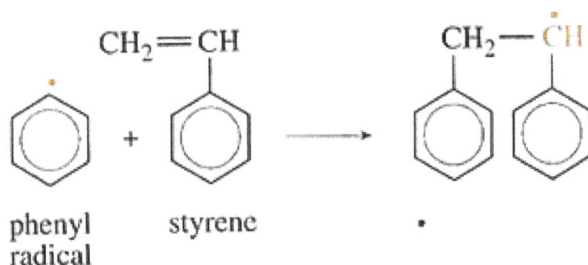

phenyl radical        styrene

The net result is that the reactants have been linked together but the product is still a radical and so is capable of attacking further monomer molecules. In each instance the attack will lead to a larger molecule but the free radical will be preserved. The reaction is referred to as free radical polymerization.

Free radicals are not the only way of initiating reactions. Charged molecules can often exert the same effect. Ethyllithium, for example, is a relatively unstable molecule which can dissociate to form an ion pair:

$$C_2H_5 Li \rightarrow C_2H_5^- + Li^+$$

Styrene can also be polymerized by this compound:

This mechanism is called anionic polymerization.

## Propagation

Once a small number of chains have been started, propagation involves successive addition of monomer units to achieve chain growth. At each step the free radical is regenerated as it reacts with the double bond. So in the case of styrene the propagation step is

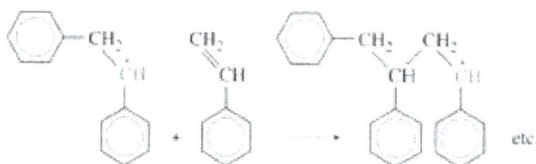

The free radical can also add on in a different way to produce

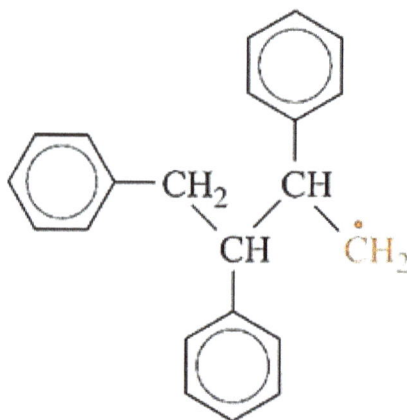

but this process happens only rarely since the free radical is less stable than in the first case.

The junction that is formed normally is known as a head-to-tail link while the abnormal link is head-to-head. The effect is limited to about 1 per cent of the total number of monomer links in normal polystyrene, but it is important because the head-to-head links are weaknesses in the chain. Since they are of higher energy, thermal degradation can start at these defective junctions.

## Termination and Transfer

There are basically three ways in which chains terminate.

The first is known as coupling and occurs when two free radicals join together. This can be represented by the general equation

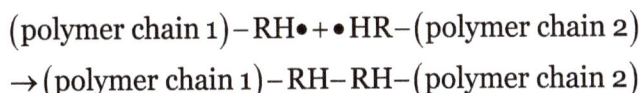

$$(\text{polymer chain 1}) - RH\bullet + \bullet HR - (\text{polymer chain 2})$$
$$\rightarrow (\text{polymer chain 1}) - RH - RH - (\text{polymer chain 2})$$

Such a mechanism significantly increases molecular mass, if it results in two polymer chains joining. This is the main mechanism which terminates the polymerization of styrene.

An alternative mechanism that may occur when two radicals interact is known as disproportionation. In this case, one molecule abstracts a hydrogen atom from the other and the other molecule forms a double bond

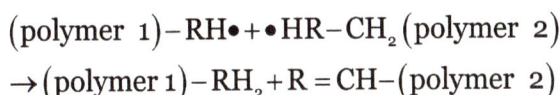

$$(\text{polymer 1}) - RH\bullet + \bullet HR - CH_2 (\text{polymer 2})$$
$$\rightarrow (\text{polymer 1}) - RH_2 + R = CH - (\text{polymer 2})$$

Disproportionation has no effect on molecular mass. Poly(methyl methacrylate) (PMMA) terminates by a mixture of coupling and disproportionation.

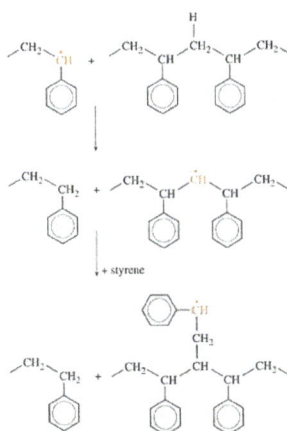

Figure: Formation of branched polymer in free radical polystyrene. The growing chain end abstracts a hydrogen atom from neighbouring polymer (top) and (middle), which then propagates with styrene monomer to form a single branch (bottom)

The third method of termination is chain transfer in which a radical abstracts a hydrogen atom from a neighbouring molecule. In the case of polystyrene the effect will be as shown in Figure, where (a) shows the situation before the interaction and (b) shows the structures after chain transfer in which the radical is transferred to one of the mid-chain carbon atoms. The new radical may now attack further styrene (Figure(c)) but, because it is not on the end of the chain, side branching occurs.

A similar mechanism accounts for the side branches in LDPE where it is a more important mode of termination than in polystyrene. Transfer to monomer; initiator or solvent (if present) can also occur in free radical polymerization, and effectively increases the dispersion of the molecular mass of the final polymer.

If termination is simply by disproportionation, then

$$n = \frac{K[M]}{[I]^{1/2}}$$

where $K$ is a constant, [M] the concentration of monomer, [I] the concentration of initiator, e.g. peroxide, and $n$ the degree of polymerization. The square root arises because two free radicals react together during termination. If termination is by coupling there will be an extra factor of two in the constant compared to disproportionation. So the degree of polymerization or molecular mass can be controlled by varying monomer concentration – for example, by conducting the reaction in solvent – or by varying initiator concentration.

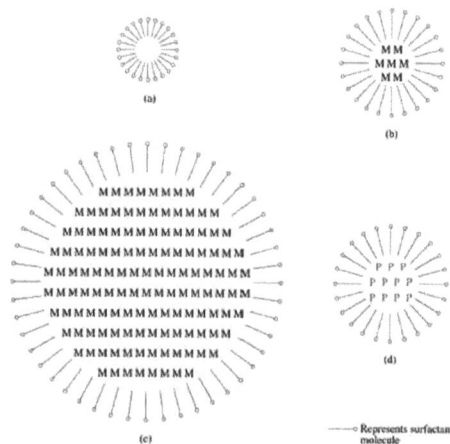

Figure: Emulsion polymerization of vinyl monomers is conducted in water to aid heat dissipation. Soap micelles (a) swell with monomer (b) which migrates from monomer droplets stabilised by soap molecules (c). The process yields a polymer latex (d) which can be used directly or reduced to bulk polymer

Controlling polymerizations on an industrial scale is of critical importance for molecular mass, and hence the processability and physical properties of the polymer, and one of the most important variables is the temperature of reaction. All polymerizations are exothermic (heat is liberated due to bond formation) and the heat must be conducted away to maintain a uniform reaction temperature. This is much more easily achieved when an inert solvent is used. Another method very commonly used industrially is to emulsify the monomer with a soap and conduct the reaction in water – so-called emulsion polymerization Figure. Since control of molecular mass is so vital, extra aids are used industrially in addition to varying monomer and initiator concentrations. Reactions are 'short stopped' before all monomer is consumed by adding a specific chemical which reacts with free radicals, stopping them dead. Other chemicals can be added to induce transfer reactions, so controlling molecular mass distribution.

## Living Polymerization

A Living polymerization is defined as a chain polymerization without irreversible transfer and termination. Living Polymerizations will lead to well-defined polymers only if the following additional prerequisites are fulfilled:

- initiation is fast in comparison with propagation,

- exchange between species of different reactivities is fast in comparison with propagation,

- rate of depropagation is low in comparison with propagation

- system is sufficiently homogeneous, in sense of the availability of active centers and mixing.

If these specifications are not met, living polymerizations will produce polymers with broader polydispersities and degrees of polymerization much higher than the $\Delta [M]/[I]o$ ratio. The proportion of chains affected by transfer and termination increases with the chain length.

## Living Polymerization Process

There are four general processes occurring simultaneously during the polymerization process: chain initiation, chain propagation, chain transfer, and chain termination. In simple terms, each polymer chain starts to grow, propagates, and terminates at a certain time, and synchronizing these processes results in chains of similar length, or molecular weight, which is desirable. If the chain initiation rate is slower or comparable to propagation rate, some chains are being initiated while others are rapidly growing, resulting in longer and shorter chains. On the other hand, if initiation is much faster than the propagation, polymer chains start growing simultaneously, and grow uniformly. Now, if the chains are not terminated by any additional mechanism, the only factor defining their growth is the presence of a monomer. Once the monomer is depleted, the growth is complete, with resulting polymer chains of the same length.

Living polymerization, which has been studied for more than 70 years, can follow anionic, cationic, and radical polymerization mechanisms. Popular atom transfer radical polymerization (ATRP) and reversible addition-fragmentation chain transfer (RAFT) are examples of living radical polymerization. Living polymerization allows you to obtain precisely controlled molecular weight and narrow molecular weight distribution, as well as complex polymer architectures.

## Techniques

### Living Anionic Polymerization

As early as 1936, Karl Ziegler proposed that anionic polymerization of styrene and butadiene by consecutive addition of monomer to an alkyl lithium initiator occurred without chain transfer or termination. Twenty years later, living polymerization was demonstrated by Szwarc through the anionic polymerization of styrene in THF using sodium naphthalenide as celerator.

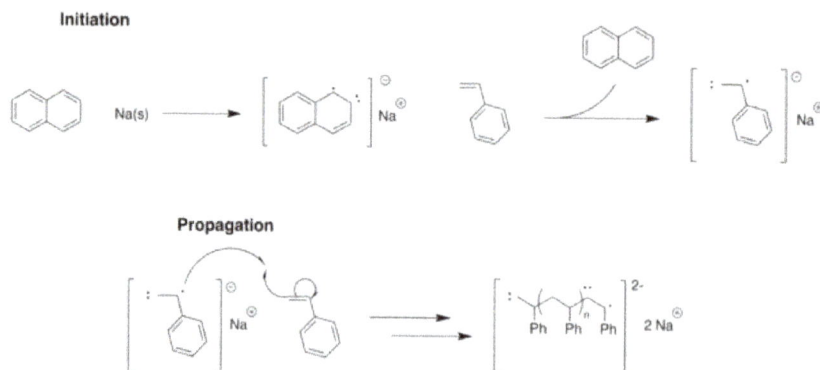

**Initiation**

**Propagation**

Here, the naphthalene anion acts as the initiator of the polymerization by activating the styrene. However, note that (with no impurities present for quenching and no solvent for chain transfer) there is no route for termination to occur. Therefore, these terminal anions will stay on the ends of the polymer until a quenching agent is introduced.

It is believed that the dianion of the polymer shown above is formed for this reaction, allowing the propagation to occur at either end of the chain. However, notice that there is no termination step (given impurities are not present to quench). This is the basis for anionic living polymerizations, where the terminal radical will exist until free monomer is available for additional propagation, or is quenched from an outside source.

### Living α-olefin Polymerization

α-olefins can be polymerized through an anionic coordination polymerization in which the metal center of the catalyst is considered the counter cation for the anionic end of the alkyl chain (through a M-R coordination). Ziegler-Natta initiators were developed in the mid-1950s and are heterogeneous initiators used in the polymerization of alpha-olefins. Not only were these initiators the first to achieve relatively high molecular weight poly(1-alkenes) (currently the most widely produced thermoplastic in the world PE(Polyethylene) and PP (Polypropylene) but the initiators were also capable of stereoselctive polymerizations which is attributed to the chiral Crystal structure of the heterogeneous initiator. Due to the importance of this discovery Ziegler and Natta were presented with the 1963 Nobel Prize in chemistry. Although the active species formed from the Ziegler-Natta initiator generally have long lifetimes (on the scale of hours or longer) the lifetimes of the propagating chains are shortened due to several chain transfer pathways (Beta-Hydride elimination and transfer to the co-initiator) and as a result are not considered living.

Metallocene initiators are considered as a type of Ziegler-Natta initiators due to the use of the two-component system consisting of a transition metal and a group I-III metal co-initiator (for example methylalumoxane (MAO) or other alkyl aluminum compounds). The metallocene initiators form homogeneous single site catalysts that were initially developed to study the impact that the catalyst structure had on the resulting polymers structure/properties; which was difficult for multi-site heterogeneous Ziegler-Natta initiators. Owing to the discrete single site on the metallocene catalyst researchers were able to tune and relate how the ancillary ligand (those not directly involved in the chemical transformations) structure and the symmetry about the chiral metal center affect the microstructure of the polymer. However, due to chain breaking reactions (mainly Beta-Hydride elimination) very few metallocene based polymerizations are known.

By tuning the steric bulk and electronic properties of the ancillary ligands and their substituents a class of initiators known as chelate initiators (or post-metallocene initiators) have been successfully used for stereospecific living polymerizations of alpha-olefins. The chelate initiators have a high potential for living polymerizations because the ancillary ligands can be designed to discourage or inhibit chain termination pathways. Chelate initiators can be further broken down based on the ancillary ligands; ansa-cyclopentyadienyl-amido initiators, alpha-diimine chelates and phenoxy-imine chelates.

- Ansa-cyclopentadienyl-amido (CpA) initiators

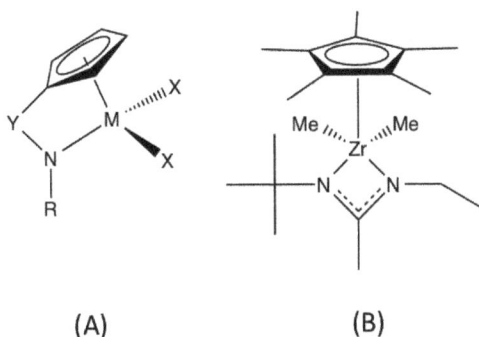

(A)                                    (B)

a.) Shows the general form of CpA initiators with one Cp ring and a coordinated Nitrogen b.) Shows the CpA initiator used in the living polymerization of 1-hexene

CpA initiators have one cyclopentadienyl substituent and one or more nitrogen substituents coordinated to the metal center (generally a Zr or Ti) (Odian). The dimethyl(pentamethylcyclopentyl)zirconium acetamidinate in figure has been used for a stereospecific living polymerization of 1-hexene at −10 deg C. The resulting poly(1-hexene) was isotactic (stereohemistry is the same between adjacent repeat units) confirmed by $^{13}$C-NMR. The multiple trials demonstrated a controllable and predictable (from catalyst to monomer ratio) $M_n$ with low Đ. The polymerization was further confirmed to be living by sequentially adding 2 portions of the monomer, the second portion was added after the first portion was already polymerized, and monitoring the Đ and $M_n$ of the chain. The resulting polymer chains complied with the predicted $M_n$ (with the total monomer concentration = portion 1 +2) and showed low Đ suggesting the chains were still active, or living, as the second portion of monomer was added.

- α-diimine chelate initiators

α-diimine chelate initiators are characterized by having a diimine chelating ancillary ligand structure and which is generally coordinated to a late transition (i.e. Ni and Pd) metal center.

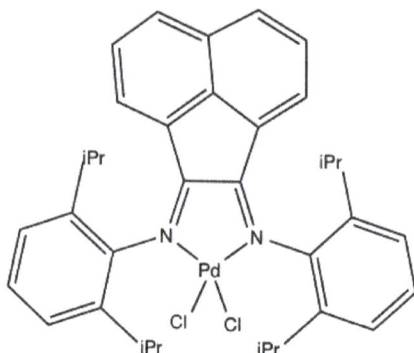

Brookhart et al. did extensive work with this class of catalysts and reported living polymerization for α-olefins and demonstrated living α-olefin carbon monoxide alternating copolymers.

- Phenoxy-imine chelates

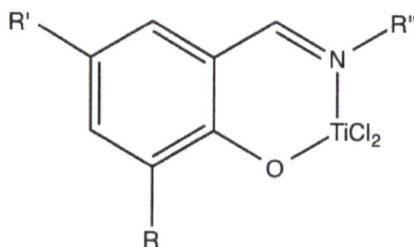

## Living Cationic Polymerization

Monomers for living cationic polymerization are electron-rich alkenes such as vinyl ethers, isobutylene, styrene, and N-vinylcarbazole. The initiators are binary systems consisting of an electrophile and a Lewis acid. The method was developed around 1980 with contributions from Higashimura, Sawamoto and Kennedy. Typically, generating a stable carbocation for a prolonged period of time is difficult, due to the possibility for the cation to be quenched by a β-protons attached to another monomer in the backbone, or in a free monomer. Therefore, a different approach is taken

(Dormant)                                    (Dormant)

$$\underset{H}{\overset{R}{\sim\!\!\!C-X}} \xrightleftharpoons{MX_n} \underset{H}{\overset{R}{\sim\!\!\!C \oplus}} MX_{n+1}^{\ominus} \xrightleftharpoons{Nu:} \underset{H}{\overset{R}{\sim\!\!\!C-Nu}}^{\oplus} MX_{n+1}^{\ominus}$$

**Propagation**
(Active)

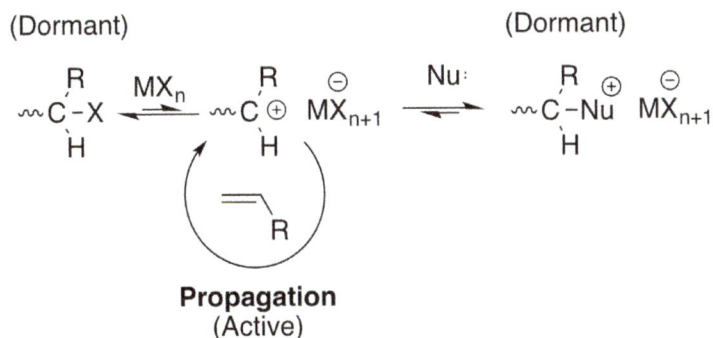

In this example, the carbocation is generated by the addition of a Lewis acid (co-initiator, along with the halogen "X" already on the polymer – see figure), which ultimately generates the carbocation in a weak equilibrium. This equilibrium heavily favors the dormant state, thus leaving little time for permanent quenching or termination by other pathways. In addition, a weak nucleophile (Nu:) can also added to reduce the concentration of active species even further, thus keeping the polymer "living". However, it is important to note that by definition, the polymers described in this example are not technically living due to the introduction of a dormant state, as termination has only been decreased, not eliminated (though this topic is still up for debate). But, they do operate similarly, and are used in similar applications to those of true living polymerizations.

## Living Ring-opening Metathesis Polymerization

Given the right reaction conditions ring-opening metathesis polymerization (ROMP) can be rendered living. The first such systems were described by Robert H. Grubbs in 1986 based on norbornene and Tebbe's reagent and in 1978 Grubbs together with Richard R. Schrock describing living polymerization with a tungsten carbene complex.

Generally, ROMP reactions involve the conversion of a cyclic olefin with significant ring-strain (>5 kcal/mol), such as cyclobutene, norbornene, cyclopentene, etc., to a polymer that also contains double bonds. The important thing to note about ring-opening metathesis polymerizations is that the double bond is usually maintained in the backbone, which can allow it to be considered "living" under the right conditions.

For a ROMP reaction to be considered "living", several guidelines must be met:

1. Fast and complete initiation of the monomer. This means that the rate at which an initiating agent activates the monomer for polymerization, must happen very quickly.

2. How many monomers make up each polymer (the degree of polymerization) must be related linearly to the amount of monomer you started with.

3. The dispersity of the polymer must be < 1.5. In other words, the distribution of how long your polymer chains are in your reaction must be very low.

With these guidelines in mind, it allows you to create a polymer that is well controlled both in content (what monomer you use) and properties of the polymer (which can be largely attributed to polymer chain length). It is important to note that living ring-opening polymerizations can be anionic *or* cationic.

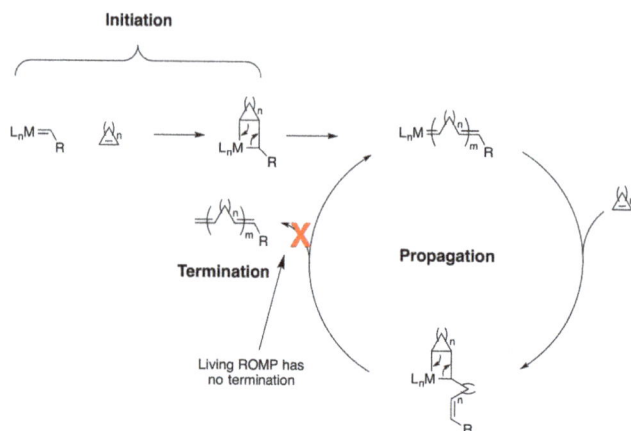

Because living polymers have had their termination ability removed, this means that once your monomer has been consumed, the addition of more monomer will result in the polymer chains continuing to grow until all of the additional monomer is consumed. This will continue until the metal catalyst at the end of the chain is intentionally removed by the addition of a quenching agent. As a result, it may potentially allow one to create a block or gradient copolymer fairly easily and accurately. This can lead to a high ability to tune the properties of the polymer to a desired application (electrical/ionic conduction, etc.)

## "Living" free Radical Polymerization

Starting in the 1970s several new methods were discovered which allowed the development of living polymerization using free radical chemistry. These techniques involved catalytic chain transfer polymerization, iniferter mediated polymerization, stable free radical mediated polymerization (SFRP), atom transfer radical polymerization (ATRP), reversible addition-fragmentation chain transfer (RAFT) polymerization, and iodine-transfer polymerization.

In "living" radical polymerization (or controlled radical polymerization (CRP)) the chain breaking pathways are severely depressed when compared to conventional radical polymerization (RP) and CRP can display characteristics of a living polymerization. However, since chain termination is not absent, but only minimized, CRP technically does not meet the requirements imposed by IUPAC for a living polymerization

There are two general strategies employed in CRP to suppress chain breaking reactions and promote fast initiation relative to propagation. Both strategies are based on developing a dynamic equilibrium amongst an active propagating radical and a dormant species.

The first strategy involves a reversible trapping mechanism in which the propagating radical undergoes an activation/deactivation (i.e. Atom-transfer radical-polymerization) process with a species X. The species X is a persistent radical, or a species that can generate a stable radical, that cannot terminate with itself or propagate but can only reversibly "terminate" with the propagating radical (from the propagating polymer chain)P*. P* is a radical species that an propagate ($k_p$) and irreversibly terminate ($k_t$) with another P*. X is normally a nitroxide (i.e. TEMPO used in Nitroxide Mediated Radical Polymerization) or an organometallic species. The dormant species ($P_n$-X) can be activated to regenerate the active propagating species (P*) spontaneously, thermally, using a catalyst and optically.

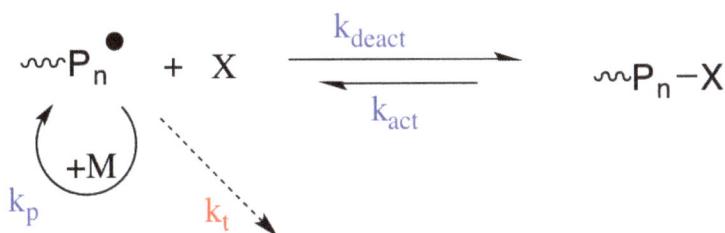

The second strategy is based on a degenerative transfer (DT) of the propagating radical between transfer agent that acts as a dormant species (i.e. Reversible addition–fragmentation chain-transfer polymerization). The DT based CRP's follow the conventional kinetics of radical polymerization, that is slow initiation and fast termination, but the transfer agent (Pm-X or Pn-X) is present in a much higher concentration compared to the radical initiator.The propagating radical species undergoes a thermally neutral exchange with the dormant transfer agent through atom transfer, group transfer or addition fragment chemistry.

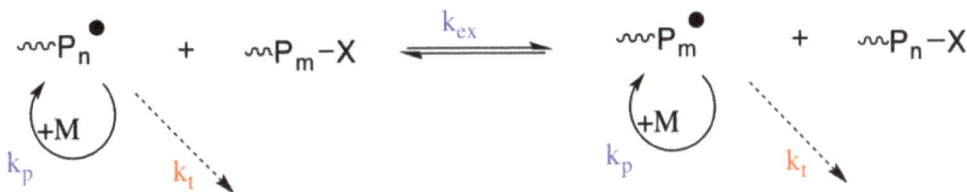

## Living Chain-growth Polycondensations

Chain growth polycondensation polymerizations were initially developed under the premise that a change in substituent effects of the polymer, relative to the monomer, causes the polymers end group to be more reactive this has been referred to as "reactive intermediate polycondensation". The essential result is monomers preferentially react with the activated polymer end groups over reactions with other monomers. This preferred reactivity is the fundamental difference when categorizing a polymerization mechanism as chain-growth as opposed to step-growth in which the monomer and polymer chain end group have equal reactivity (the reactivity is uncontrolled). Several strategies were employed to minimize monomer-monomer reactions (or self-condensation) and polymerizations with low Đ and controllable Mn have been attained by this mechanism for small molecular weight polymers. However, for high molecular weight polymer chains (i.e. small initiator to monomer ratio) the Mn is not easily to controlled, for some monomers, since self-condensation between monomers occurred more frequently due to the low propagating species concentration.

## Catalyst-transfer Polycondensation

Catalyst transfer polycondensation (CTP) is a chain-growth polycondensation mechanism in which the monomers do not directly react with one another and instead the monomer will only react with the polymer end group through a catalyst-mediated mechanism. The general process consists of the catalyst activating the polymer end group followed by a reaction of the end group with a 2nd incoming monomer. The catalyst is then transferred to the elongated chain while activating the end group (as shown below).

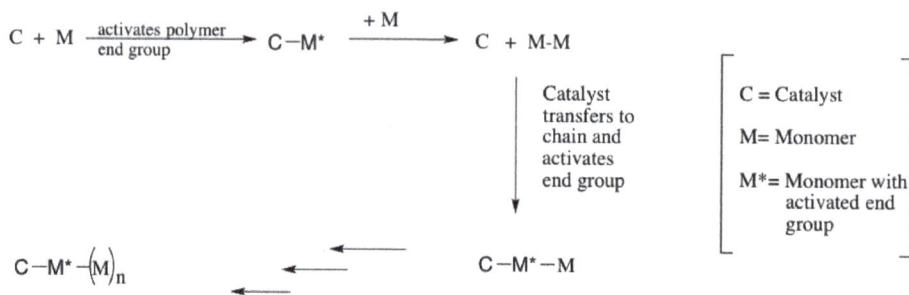

Catalyst transfer polycondensation allows for the living polymerization of π-conjugated polymers and was discovered by Tsutomu Yokozawa in 2004 and Richard McCullough. In CTP the propagation step is based on organic cross coupling reactions (i.e. Kumada coupling, Sonogashira coupling, Negishi coupling) top form carbon carbon bonds between difunctional monomers. When Yokozawa and McCullough independently discovered the polymerization using a metal catalyst to couple a Grignard reagent with an organohalide making a new carbon-carbon bond. The mechanism below shows the formation of poly(3-alkylthiophene) using a Ni initiator ($L_n$ can be 1,3-Bis(diphenylphosphino)propane (dppp)) and is similar to the conventional mechanism for Kumada coupling involving an oxidative addition, a transmetalation and a reductive elimination step. However, there is a key difference, following reductive elimination in CTP, an associative complex is formed (which has been supported by intra-/intermolecular oxidative addition competition experiments ) and the subsequent oxidative addition occurs between the metal center and the associated chain (an intramolecular pathway). Whereas in a coupling reaction the newly formed alkyl/aryl compound diffuses away and the subsequent oxidative addition occurs between an incoming Ar-Br bond and the metal center. The associative complex is essential to for polymerization to occur in a living fashion since it allows the metal to undergo a preferred intramolecular oxidative addition and remain with a single propagating chain (consistent with chain-growth mechanism), as opposed to an intermolecular oxidative addition with other monomers present in the solution (consistent with a step-growth, non-living, mechanism). The monomer scope of CTP has been increasing since its discovery and has included poly(phenylene)s, poly(fluorine)s, poly(selenophene)s and poly(pyrrole)s.

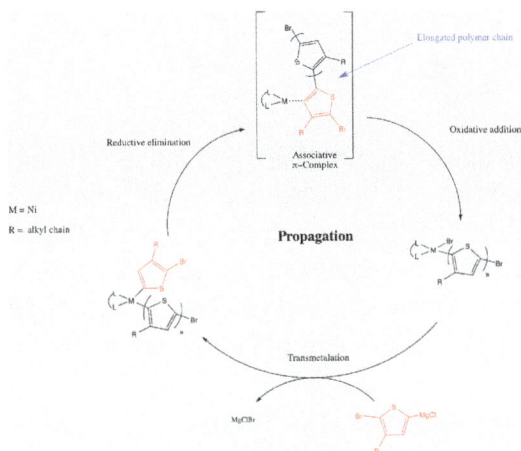

## Living Group-transfer Polymerization

Group-transfer polymerization also has characteristics of living polymerization. It is applied to alkylated methacrylate monomers and the initiator is a silyl ketene acetal. New monomer adds to the initiator and to the active growing chain in a Michael reaction. With each addition of a monomer group the trimethylsilyl group is transferred to the end of the chain. The active chain-end is not ionic as in anionic or cationic polymeriation but is covalent. The reaction can be catalysed by bifluorides and bioxyanions such as *tris(dialkylamino)sulfonium bifluoride* or *tetrabutyl ammonium bibenzoate*. The method was discovered in 1983 by O.W. Webster and the name first suggested by Barry Trost.

## Applications

Living polymerizations can be (and in some cases are) used industrially for many different applications. They can range from self-healing materials for space equipment to the easy design of copolymers for ion-exchange membranes in fuel cells, nanoscale lithography, etc.. While living polymerizations are still not widely used industrially, the field is rapidly growing, as well as the list of practical applications.

## Self-healing Materials

Self-healing materials are materials in which repair, or "heal", themselves upon damage from an external force through the use of living polymers. For example, if a crack forms in the material, it proceeds to repair the crack and restore itself to its original, undamaged form. It achieves this by incorporating monomer-containing beads into a material made of a living polymer (with a terminally active chain). This has been achieved recently using a polyurethane derivative, with beads of monomer embedded in the material that become opened upon cracking of the material.

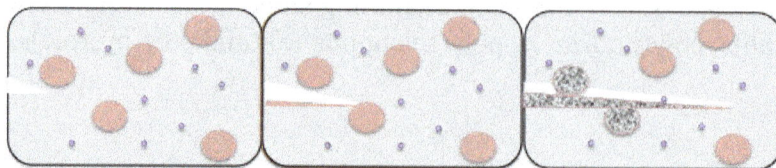

The polymer that makes up the material is designed as a living polymer, with reactive terminal end-groups that bind to the freshly provided monomer upon damage to the microbeads. This addition of monomer to the polymer chain increases the polymer chain to a length that fills the once open crack, in essence reconnecting all of the pieces back into one. According to Odriozola and coworkers, this application is originally designed for space equipment (in the event of debris damaging the equipment).

## Copolymer Synthesis and Applications

Copolymers are polymers consisting of multiple different monomer species, and can be arranged in various orders, three of which are seen in the figure below.

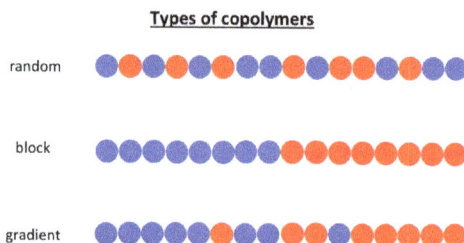

Types of copolymers

While there exist others (alternating copolymers, graft copolymers, and stereoblock copolymers), these three are more common in the scientific literature. In addition, block copolymers can exist as many types, including triblock (A-B-A), alternating block (A-B-A-B-A-B), etc.

Of these three types, block and gradient copolymers are commonly synthesized through living polymerizations, due to the ease of control living polymerization provides. Copolymers are highly desired due to the increased flexibility of properties a polymer can have compared to their homopolymer counterparts. The synthetic techniques used range from ROMP to generic anionic or cationic living polymerizations.

Copolymers, due to their unique tunability of properties, can have a wide range of applications. One example (of many) is nano-scale lithography using block copolymers. One used frequently is a block copolymer made of polystyrene and poly(methyl methacrylate) (abbreviated PS-*b*-PMMA). This copolymer, upon proper thermal and processing conditions, can form cylinders on the order of a few tens of nanometers in diameter of PMMA, surrounded by a PS matrix. These cylinders can then be etched away under high exposure to UV light and acetic acid, leaving a porous PS matrix.

The unique property of this material is that the size of the pores (or the size of the PMMA cylinders) can be easily tuned by the ratio of PS to PMMA in the synthesis of the copolymer. This can be easily tuned due to the easy control given by living polymerization reactions, thus making this technique highly desired for various nanoscale patterning of different materials for applications to catalysis, electronics, etc.

## Cationic Polymerization

Many vinyl monomers readily polymerize in the presence of very small amounts of catalyst of the type used in Friedel-Crafts reactions. Examples of effective catalysts are $AlCl_3$, $AlBr_3$, $BF_3$, $TiCl_4$, $SnCl_4$, and in some cases strong acids such as $H_2SO_4$. All these catalysts are examples of Lewis acids with strong electron-acceptor capability. They usually require a co-catalyst, namely a Lewis base such as Water, acetic acid or alcohol:

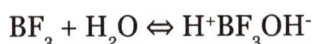

$$BF_3 + H_2O \Leftrightarrow H^+BF_3OH^-$$

Monomers that polymerize in the presence of these catalysts include isobutylene, styrene, alpha-methylstyrene, butadiene, vinyl alkyl ethers and many other monomers having electron-donating substituents that enhance the electron-sharing ability of the double bond of the vinyl monomers.

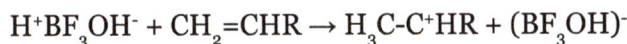

$$H^+BF_3OH^- + CH_2=CHR \rightarrow H_3C-C^+HR + (BF_3OH)^-$$

They all can be readily polymerized to high-molecular weight polymers.

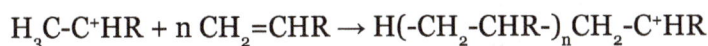

$$H_3C-C^+HR + n\ CH_2=CHR \rightarrow H(-CH_2-CHR-)_nCH_2-C^+HR$$

However, some other monomers, such as propylene and other olefins, reach only low to medium molecular weights when polymerized with strong Lewis acids.

The cationic polymerization usually proceeds at high rates both at high and (very) low temperatures. For this reason, a uniform and constant reaction condition cannot be maintained during polymerization. For example, isobutylene at -100C in the presence of a strong Lewis acid polymerizes to high molecular weight polybutylene within a fraction of a second. To prevent excessive rise in temperature in the reaction vessel, an "internal refrigerant" is usually added to the mixture that dissipates the heat by evaporation of a portion of the liquid.

Both the rate of reaction and the molecular weight decreases with increasing temperature. For this reason, low temperatures are usually preferred. In fact, molecular weights obtained at room temperature are often much lower than those achieved by free-radical polymerization.

# Synthesis

## Initiation

Initiation is the first step in cationic polymerization. During initiation, a carbenium ion is generated from which the polymer chain is made. The counterion should be non-nu-cleophilic, otherwise the reaction is terminated instantaneously. There are a variety of initiators available for cationic polymerization, and some of them require a coinitiator to generate the needed cationic species.

## Classical Protonic Acids

Strong protic acids can be used to form a cationic initiating species. High concentra-tions of the acid are needed in order to produce sufficient quantities of the cationic species. The counterion ($A^-$) produced must be weakly nucleophilic so as to prevent early termination due to combination with the protonated olefin. Common acids used are phosphoric, sulfuric, fluro-, and triflic acids. Only low molecular weight polymers are formed with these initiators.

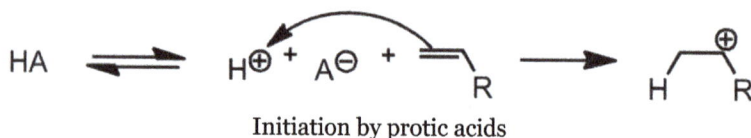

Initiation by protic acids

## Lewis acids/Friedel-Crafts catalysts

Lewis acids are the most common compounds used for initiation of cationic polymer-ization. The more popular Lewis acids are $SnCl_4$, $AlCl_3$, $BF_3$, and $TiCl_4$. Although these Lewis acids alone are able to induce polymerization, the reaction occurs much faster with a suitable cation source. The cation source can be water, alcohols, or even a car-bocation donor such as an ester or an anhydride. In these systems the Lewis acid is referred to as a coinitiator while the cation source is the initiator. Upon reaction of the initiator with the coinitiator, an intermediate complex is formed which then goes on to react with the monomer unit. The counterion produced by the initiator-coinitiator complex is less nucleophilic than that of the Brønsted acid $A^-$ counterion. Halogens, such as chlorine and bromine, can also initiate cationic polymerization upon addition of the more active Lewis acids.

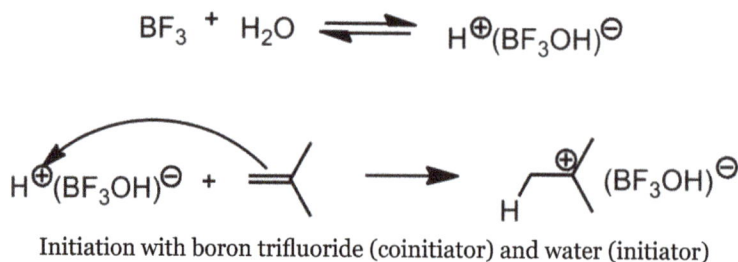

Initiation with boron trifluoride (coinitiator) and water (initiator)

## Carbenium ion salts

Stable carbenium ions are used to initiate chain growth of only the most reactive olefins and are known to give well defined structures. These initiators are most often used in kinetic studies due to the ease of being able to measure the disappearance of the carbenium ion absorbance. Common carbenium ions are trityl and tropylium cations.

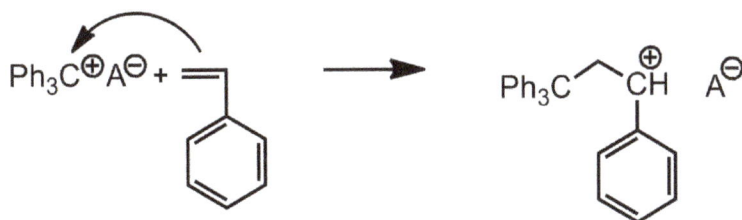

Initiation with trityl carbenium ion

## Ionizing Radiation

Ionizing radiation can form a radical-cation pair that can then react with a monomer to start cationic polymerization. Control of the radical-cation pairs are difficult and often depend on the monomer and reaction conditions. Formation of radical and anionic species are often observed.

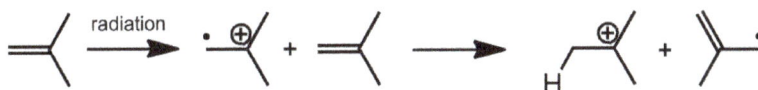

Initiation using ionizing radiation

## Propagation

Propagation proceeds via addition of monomer to the active species, i.e. the carbenium ion. The monomer is added to the growing chain in a head-to-tail fashion; in the process, the cationic end group is regenerated to allow for the next round of monomer addition.

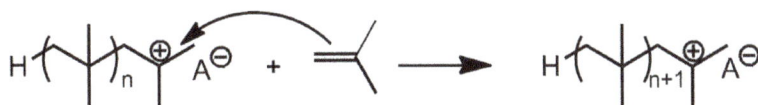

General propagation pathway

## Effect of Temperature

The temperature of the reaction has an effect on the rate of propagation. The overall activation energy for the polymerization ($E$) is based upon the activation energies for the initiation ($E_i$), propagation ($E_p$), and termination ($E_t$) steps:

$$E = E_i + E_p - E_t$$

Generally, $E_t$ is larger than the sum of $E_i$ and $E_p$, meaning the overall activation energy is negative. When this is the case, a decrease in temperature leads to an increase in the rate of propagation. The converse is true when the overall activation energy is positive.

Chain length is also affected by temperature. Low reaction temperatures, in the range of 170–190 K, are preferred for producing longer chains. This comes as a result of the activation energy for termination and other side reactions being larger than the activation energy for propagation. As the temperature is raised, the energy barrier for the termination reaction is overcome, causing shorter chains to be produced during the polymerization process.

## Effect of Solvent and Counterion

The solvent and the counterion (the gegen ion) have a significant effect on the rate of propagation. The counterion and the carbenium ion can have different associations according to intimate ion pair theory; ranging from a covalent bond, tight ion pair (unseparated), solvent-separated ion pair (partially separated), and free ions (completely dissociated).

~~~RX covalent      ~~~R$^+$X$^-$ tight ion pair      ~~~R$^+$/X$^-$ solvent-separated ion pair      ~~~R$^+$+X$^-$ free ions

Range of associations between the carbenium ion ($R^+$) and gegen ion ($X^-$)

The association is strongest as a covalent bond and weakest when the pair exists as free ions. In cationic polymerization, the ions tend to be in equilibrium between an ion pair (either tight or solvent-separated) and free ions. The more polar the solvent used in the reaction, the better the solvation and separation of the ions. Since free ions are more reactive than ion pairs, the rate of propagation is faster in more polar solvents.

The size of the counterion is also a factor. A smaller counterion, with a higher charge density, will have stronger electrostatic interactions with the carbenium ion than will a larger counterion which has a lower charge density. Further, a smaller counterion is more easily solvated by a polar solvent than a counterion with low charge density. The result is increased propagation rate with increased solvating capability of the solvent.

## Termination

Termination generally occurs via unimolecular rearrangement with the counterion. In this process, an anionic fragment of the counterion combines with the propagating chain end. This not only inactivates the growing chain, but it also terminates the kinetic chain by reducing the concentration of the initiator-coinitiator complex.

Termination by combination with an anionic fragment from the counterion

## Chain Transfer

Chain transfer can take place in two ways. One method of chain transfer is hydrogen abstraction from the active chain end to the counterion. In this process, the growing chain is terminated, but the initiator-coinitiator complex is regenerated to initiate more chains.

Chain transfer by hydrogen abstraction to the counterion

The second method involves hydrogen abstraction from the active chain end to the monomer. This terminates the growing chain and also forms a new active carbenium ion-counterion complex which can continue to propagate, thus keeping the kinetic chain intact.

Chain transfer by hydrogen abstraction to the monomer

## Cationic Ring-opening Polymerization

Cationic ring-opening polymerization follows the same mechanistic steps of initiation, propagation, and termination. However, in this polymerization reaction, the monomer units are cyclic in comparison to the resulting polymer chains which are linear. The linear polymers produced can have low ceiling temperatures, hence end-capping of the polymer chains is often necessary to prevent depolymerization.

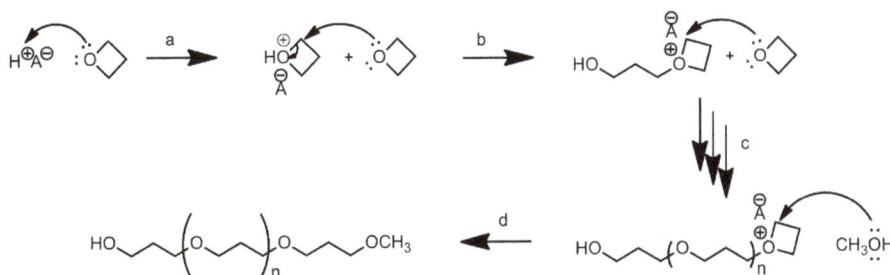

Cationic ring-opening polymerization of oxetane involving (a and b) initiation, (c) propagation, and (d) termination with methanol

## Kinetics

The rate of propagation and the degree of polymerization can be determined from an analysis of the kinetics of the polymerization. The reaction equations for initiation, propagation, termination, and chain transfer can be written in a general form:

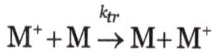

$$I{+}{+}M \xrightarrow{k_i} M^+$$

$$M^+ + M \xrightarrow{k_p} M^+$$

$$M^+ \xrightarrow{k_t} M$$

$$M^+ + M \xrightarrow{k_{tr}} M + M^+$$

In which $I^+$ is the initiator, M is the monomer, $M^+$ is the propagating center, and $k_i$, $k_p$, $k_t$, and $k_{tr}$ are the rate constants for initiation, propagation, termination, and chain transfer, respectively. For simplicity, counterions are not shown in the above reaction equations and only chain transfer to monomer is considered. The resulting rate equations are as follows, where brackets denote concentrations:

$$\text{rate(initiation)} = k_i[I^+][M]$$

$$\text{rate(propagation)} = k_p[M^+][M]$$

$$\text{rate(termination)} = k_t[M^+]$$

$$\text{rate(chain transfer)} = k_{tr}[M{+}][M]$$

Assuming steady-state conditions, i.e. the rate of initiation = rate of termination:

$$[M^+] = \frac{k_i[I^+][M]}{k_t}$$

This equation for [M$^+$] can then be used in the equation for the rate of propagation:

$$\text{rate(propagation)} = \frac{k_p k_i [M]^2 [I^+]}{k_t}$$

From this equation, it is seen that propagation rate increases with increasing monomer and initiator concentration.

The degree of polymerization, $X_n$, can be determined from the rates of propagation and termination:

$$X_n = \frac{\text{rate(propagation)}}{\text{rate(termination)}} = \frac{k_p[M]}{k_t}$$

If chain transfer rather than termination is dominant, the equation for $X_n$ becomes

$$X_n = \frac{\text{rate(propagation)}}{\text{rate(chain transfer)}} = \frac{k_p}{k_{tr}}$$

## Living Polymerization

In 1984, Higashimura and Sawamoto reported the first living cationic polymerization for alkyl vinyl ethers. This type of polymerization has allowed for the control of well-defined polymers. A key characteristic of living cationic polymerization is that termination is essentially eliminated, thus the cationic chain growth continues until all monomer is consumed.

## Commercial Applications

The largest commercial application of cationic polymerization is in the production of polyisobutylene (PIB) products which include polybutene and butyl rubber. These polymers have a variety of applications from adhesives and sealants to protective gloves and pharmaceutical stoppers. The reaction conditions for the synthesis of each type of isobutylene product vary depending on the desired molecular weight and what type(s) of monomer(s) is used. The conditions most commonly used to form low molecular weight (5–10 x $10^4$ Da) polyisobutylene are initiation with $AlCl_3$, $BF_3$, or $TiCl_4$ at a temperature range of −40 to 10 °C. These low molecular weight polyisobutylene polymers are used for caulking and as sealants. High molecular weight PIBs are synthesized at much lower temperatures of −100 to −90 °C and in a polar medium of methylene chloride. These polymers are used to make uncrosslinked rubber products and are additives for certain thermoplasts. Another characteristic of high molecular weight PIB is its low toxicity which allows it to be used as a base for chewing gum. The main chemical companies that produce polyisobutylene are Esso, ExxonMobil, and BASF.

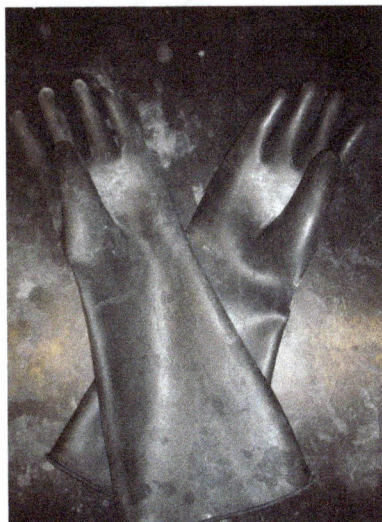

Butyl rubber gloves

Butyl rubber, in contrast to PIB, is a copolymer in which the monomers isobutylene (~98%) and isoprene (2%) are polymerized in a process similar to high molecular weight PIBs. Butyl rubber polymerization is carried out as a continuous process with $AlCl_3$ as the initiator. Its low gas permeability and good resistance to chemicals and aging make it useful for a variety of applications such as protective gloves, electrical cable insulation, and even basketballs. Large scale production of butyl rubber started during World War II, and roughly 1 billion pounds/year are produced in the U.S. today.

Polybutene is another copolymer, containing roughly 80% isobutylene and 20% other butenes (usually 1-butene). The production of these low molecular weight polymers (300–2500 Da) is done within a large range of temperatures (−45 to 80 °C) with $AlCl_3$ or $BF_3$. Depending on the molecular weight of these polymers, they can be used as adhesives, sealants, plasticizers, additives for transmission fluids, and a variety of other applications. These materials are low-cost and are made by a variety of different companies including BP Chemicals, Esso, and BASF.

Other polymers formed by cationic polymerization are homopolymers and copolymers of polyterpenes, such as pinenes (plant-derived products), that are used as tackyfiers. In the field of heterocycles, 1,3,5-trioxane is copolymerized with small amounts of ethylene oxide to form the highly crystalline polyoxymethylene plastic. Also, the homopolymerization of alkyl vinyl ethers is achieved only by cationic polymerization.

## Anionic Polymerization

Addition polymers can also be made by chain reactions that proceed through intermediates that carry either a negative or positive charge.

When the chain reaction is initiated and carried by negatively charged intermediates, the reaction is known as anionic polymerization. Like free-radical polymerizations, these chain reactions take place via chain-initiation, chain-propagation, and chain-termination steps.

## Monomer Characteristics

In order for polymerization to occur with vinyl monomers, the substituents on the double bond must be able to stabilize a negative charge. Stabilization occurs through delocalization of the negative charge. Because of the nature of the carbanion propagating center, substituents that react with bases or nucleophiles either must not be present or be protected.

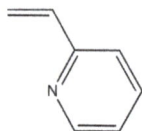

Butadiene

vinyl pyridine

Examples of vinyl monomers.

Vinyl monomers with substituents that stabilize the negative charge through charge delocalization, undergo polymerization without termination or chain transfer. These monomers include styrene, dienes, methacrylate, vinyl pyridine, aldehydes, epoxide, episulfide, cyclic siloxane, and lactones. Polar monomers, using controlled conditions and low temperatures, can undergo anionic polymerization. However, at higher temperatures they do not produce living stable, carbanionic chain ends because their polar substituents can undergo side reactions with both initiators and propagating chain centers. The effects of counterion, solvent, temperature, Lewis base additives, and inorganic solvents have been investigated to increase the potential of anionic polymerizations of polar monomers. Polar monomers include acrylonitrile, cyanoacrylate, propylene oxide, vinyl ketone, acrolein, vinyl sulfone, vinyl sulfoxide, vinyl silane and isocyanate.

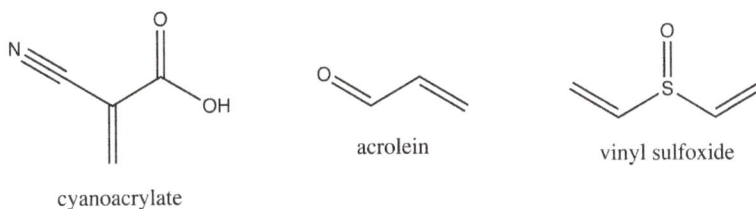

cyanoacrylate                     acrolein                     vinyl sulfoxide

Examples of polar monomers.

## Solvent

The solvent used in anionic addition polymerizations are determined by the reactivity of both the initiator and carbanion of the propagating chain end. Anionic species with low reactivity, such as heterocyclic monomers, can use a wide range of solvents.

## Initiation

The reactivity of initiators used in anionic polymerization should be similar to that of the monomer that is the propagating species. The pKa values for the conjugate acids of the carbanions formed from monomers can be used to deduce the reactivity of the monomer. The least reactive monomers have the largest pKa values for their corresponding conjugate acid and thus, require the most reactive initiator. Two main initiation pathways involve electron transfer (through alkali metals) and strong anions.

## Initiation by Electron Transfer

Szwarc and coworkers studied the initiation of polymerization through the use of aromatic radical-anions such as sodium naphthenate. In this reaction, an electron is transferred from the alkali metal to naphthalene. Polar solvents are necessary for this type of initiation both for stability of the anion-radical and to solvate the cation species formed. The anion-radical can then transfer an electron to the monomer.

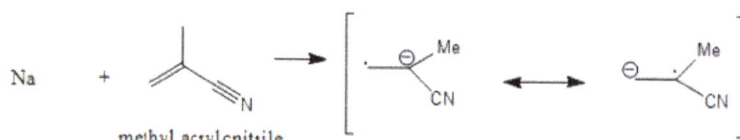

methyl acrylcnitrile

Initiation through electron transfer.

Initiation can also involve the transfer of an electron from the alkali metal to the monomer to form an anion-radical. Initiation occurs on the surface of the metal, with the reversible transfer of an electron to the adsorbed monomer.

## Initiation by Strong Anions

Nucleophilic initiators include covalent or ionic metal amides, alkoxides, hydroxides, cyanides, phosphines, amines and organometallic compounds (alkyllithium compounds and Grignard reagents). The initiation process involves the addition of a neutral (B:) or negative (B:-) nucleophile to the monomer.

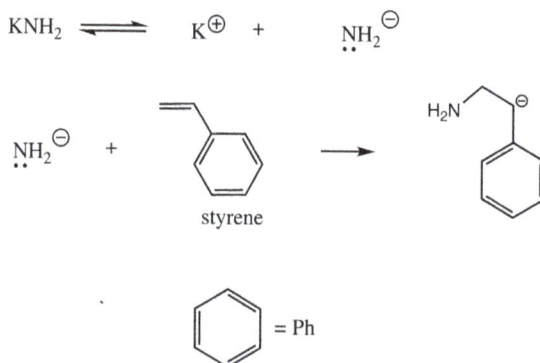

styrene

= Ph

Initiation through strong anion.

The most commercially useful of these initiators has been the alkyllithium initiators. They are primarily used for the polymerization of styrenes and dienes.

Monomers activated by strong electronegative groups may be initiated even by weak anionic or neutral nucleophiles (i.e. amines, phosphines). Most prominent example is the curing of cyanoacrylate, which constitutes the basis for superglue. Here, only traces of basic impurities are sufficient to induce an anionic addition polymerization or zwitterionic addition polymerization, respectively.

## Propagation

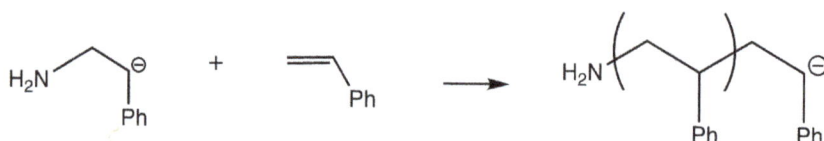

Propagation of an anionic addition polymerization.

Propagation in anionic addition polymerization results in the complete consumption of monomer. It is very fast and occurs at low temperatures. This is due to the anion not being very stable, the speed of the reaction as well as that heat is released during the reaction. The stability can be greatly enhanced by reducing the temperatures to near $0°C$. The propagation rates are generally fairly high compared to the decay reaction, so the overall polymerization rates is generally not affected.

## Termination

Anionic addition polymerizations have no formal termination pathways because proton transfer from solvent or other positive species does not occur. However, termination can occur through unintentional quenching due to trace impurities. This includes trace amounts of oxygen, carbon dioxide or water. Intentional termination can occur through the addition of water or alcohol. Another method of termination, chain transfer, can occur when an agent can act as a Brønsted acid. In this case, the pKa value of the agent is similar to the conjugate acid of the propagating carbanionic chain end. Spontaneous termination occurs because the concentration of carbanion centers decay over time and eventually results in hydride elimination. Polar monomers are more reactive because they are stabilized by their polar substituents. These polar substituents can react with nucleophiles which results in termination as well as side reactions that compete with both initiation and propagation.

## Living Anionic Polymerization

Living polymerization was demonstrated by Szwarc and co workers in 1956. Their initial work was based on the polymerization of styrene and dienes. One of the remarkable features of living anionic polymerization is that the mechanism involves no formal termination step. In the absence of impurities, the carbanion would still be active and capable of adding another monomer. The chains will remain active indefinitely unless there is inadvertent or deliberate termination or chain transfer.

## Kinetics

The kinetics of anionic addition polymerization depend on whether or not a termination pathway occurs.

### Kinetics of Living Anionic Addition Polymerization

In general, the reaction mechanism for living anionic addition polymerization are as follows:

$$I^- + M \xrightarrow{k_{init}} M^-$$

$$M^- + M \xrightarrow{k_{prop}} M^-$$

where I = initiator, $k_{init}$ = the initiation reaction rate constant, M = monomer, M⁻= propagating species, and $k_{prop}$ = the propagation reaction rate constant.

As most polymerizations of this type do not have a termination pathway, the rate of polymerization is the rate of propagation:

$$\text{rate(prop)} = k_p[\text{M}^-][\text{M}]$$

where $k_p$ is the rate of constant of propagation, [M⁻] is the total concentration of propagating centers, and [M] is the concentration of monomer. Since there is no termination pathway in living polymerizations, the concentration of propagating centers is equal to the concentration of initiator. Thus,

$$\text{rate(prop)} = k_p[\text{I}][\text{M}]$$

The degree of polymerization, $X_n$ is also affected by no termination pathway. It is the ratio of concentration of reacted monomer ($[\text{M}]_o$) to initiator($[\text{I}]_o$) times the percent conversion $p$. In this case, the chain length (v) is equal to $X_n$.

$$v = \frac{[\text{M}]_o}{[\text{I}]_o} \rho$$

When conversion, $p = 1$ (100% conversion), chain length is simply the ratio of reacted monomer to initiator.

$$v = \frac{[\text{M}]_o}{[\text{I}]_o}$$

## Kinetics: termination Due to Impurities

When termination occurs due to impurities, the impurities must be taken into account in determining the reaction rate. The reaction mechanisms would begin the same as that of a living anionic addition (initiation and propagation). However, there would now be a termination step to account for the effect of the impurities on the reaction.

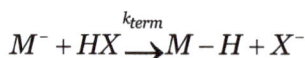

$$M^- + HX \xrightarrow{k_{term}} M-H + X^-$$

where M⁻= propagating species, HX = impurity and $k_{term}$ = the termination reaction rate constant.

Using the steady-state approximation, the rate of propagation becomes

$$\text{rate(prop)} = \frac{k_{init}k_{prop}[\text{I}][\text{M}]^2}{k_{term}[\text{H-X}]}$$

Since

$$v = \frac{\text{rate(prop)}}{\text{rate(term)}} = \frac{k_{prop}[M]}{k_{term}[\text{H-X}]}$$

Thus chain length and rate of propagation are negatively impacted by the presence of impurities in the reaction.

## Ring-opening Polymerization

A ring-opening polymerization (ROP) is another form of chain-growth polymerization in which the terminal end group of a polymer chain acts as a reactive center where further cyclic monomers can be added by ring-opening and additon of the broken bond. Typical cyclic monomers that can be polymerized via ROP are di-functional monomers that carry two different reactive groups like one amine or alcohol and one carboxylic acid that have undergone a cyclization reaction. Two examples are caprolactam and caprolactone:

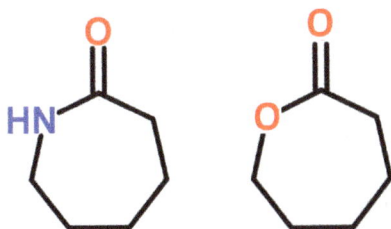

To polymerize these moieties, one of the rings has to open prior polymerization. This can be achieved, for example, by adding a small amount of a nucleophilic reagent (Lewis base) as an initiator. This reaction is called *anionic ring-opening polymerization* (AROP):

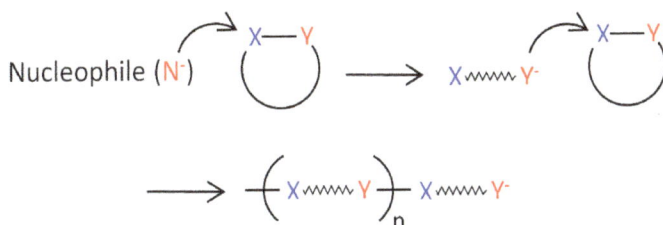

Two well known thermoplastic polymers that can be synthesized via anionic ring-opening polymerization are polycaprolactam (Nylon 6) and polycaprolactone (PCL):

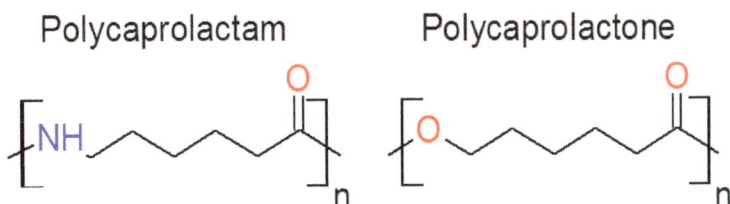

Most monomers that undergo AROP contain polar bonds like ester, amide, carbonate, urethane, epoxide, and phosphate which polymerize to the corresponding polyester, polyamide, polycarbonate, polyurethane, polyepoxide, and polyphosphate.

*Cationic ring-opening polymerization* (CROP) is also possible. In this case, a small amount of an electrophilic reagent (Lewis acid) is added to the monomer to initiate polymerization. However, not all cyclic monomers containing an heteroatom undergo CROP. Whether and how readily a cyclic monomer undergoes CROP depends on the ring size, to be more specific, on the ring strain. Cyclic monomers with small or no ring strain will not polymerize whereas small rings with greater ring strain like 4, 6, and 7-membered rings of cyclic esters, polymerize readily through CROP.

Some examples of cyclic monomers that polymerize through anionic or cationic ring-opening polymerization include cyclic ethers, lactones, lactams, and epoxides.

Ring-opening polymerization can also proceed via *free radical polymerization*. The introduction of an oxygen into the ring will usually promote free radical ring-opening polymerization, because the resulting carbon–oxygen double bond is much more stable than a carbon-carbon double bond. Thus, cyclic hetero monomers that carry a vinyl side group like cyclic ketene acetals, cyclic ketene aminals, cyclic vinyl ethers, and unsaturated spiro ortho esters will readily undergo free radical ring-opening polymerization. Copolymerization of these monomers with a wide variety of vinyl monomers will introduce ester, amide, keto or carbonate groups into the backbone, which results in functionally terminated oligomers.

## References

- Stevens, Malcolm P. (1999). Polymer Chemistry: An Introduction. New York: Oxford University Press. ISBN 0-19-512444-8

- Islamova, R. M.; Puzin, Y. I.; Kraikin, V. A.; Fatykhov, A. A.; Dzhemilev, U. M. (2006). "Controlling the Polymerization of Methyl Methacrylate with Ternary Initiating Systems". Russian Journal of Applied Chemistry. 79 (9): 1509–1513. doi:10.1134/S1070427206090229

- Polymerization: adhesiveandglue.com, Retrieved 02 July 2018

- "Stable Free Radical Polymerization". Xerox Corp. 2010. Archived from the original on 28 November 2003. Retrieved 10 March 2010

- Cowie, J. M. G. (1991). Polymers: Chemistry and Physics of Modern Materials (2nd ed.). Blackie (USA: Chapman & Hall). pp. 58–60. ISBN 0-216-92980-6

- Macro-reactions-radical, Organic-chemistry-3: tut.fi, Retrieved 20 March 2018

- "Polymer Synthesis". Case Western Reserve University. 2009. Archived from the original on 7 February 2010. Retrieved 10 March 2010

- Alfrey, Turner; Price, Charles C. (1947). "Relative reactivities in vinyl copolymerization". Journal of Polymer Science. 2 (1): 101–106. Bibcode:1947JPoSc...2..101A. doi:10.1002/pol.1947.120020112

- Living-polymerization-and-molecular-weight: polymersolutions.com, Retrieved 22 May 2018

- Odian, George (27 Feb 2004). "8". Principles of Polymerization (4th ed.). John Wiley & Sons, Inc. p. 633. ISBN 978-0-471-27400-1.

# Polymer Science and Architecture

The sub-field of materials science that deals with polymers is known as polymer science. Polymer architecture studies the way branching leads to a deviation from a linear polymer chain. The aim of this chapter is to explore the important aspects of polymer science and architecture such as branched polymers, two-dimensional polymers, etc. for a comprehensive understanding of the subject matter.

## Polymer Science

Polymer science is an interdisciplinary area comprised of chemical, physical, engineering, processing, and theoretical fields of science. Its objective is to provide the basis for the creation and characterization of polymeric materials and an understanding for the relationship between structure and property.

### Future Scope of Polymer Science

The main concerns for humans in the future will be energy & resources, food, health, mobility & infrastructure and communication. There is no doubt that polymers will play a key role in finding successful ways in handling these challenges. Polymers will be the material of the new millennium and the production of polymeric parts i.e. green, sustainable, energy-efficient, high quality, low-priced, etc. will assure the accessibility of the finest solutions round the globe. Synthetic polymers have since a long time played a relatively important role in present-day medicinal practice. Many devices in medicine and even some artificial organs are constructed with success from synthetic polymers. It is possible that synthetic polymers may play an important role in future pharmacy, too. Polymer science can be applied to save energy and improve renewable energy technologies.

Biopolymers could especially increment as more solid adaptations are produced, and the cost to fabricate these bio-plastics keeps ongoing fall. Bio-plastics can supplant routine plastics in the field of their applications likewise and can be utilized as a part of various areas, for example, sustenance bundling, plastic plates, mugs, cutlery, plastic stockpiling packs and in this way can help in making environment economical.

In areas of applications of plastics materials, a well-known long standing example is electrical industries have led to increasing acceptance of plastics for plugs, sockets, wire and cable insulations and for housing electrical and electronic equipment. The major polymer targeting industries of the present day life includes Ceramic industries, in stem cell biology and Regenerative Medicine, packaging industries, in retorting method used for food processing industries in automotive industries, in aerospace industries and in electrical and electronic industries.

- Polymers in Stem Cell Biology

- Self-Healing and Reprocess-able Polymer Systems

- Smart Polymers

- Green Synthesis of Functional Materials

- In Gene Delivery Systems

- Ceramic Industry

- Biopolymers in Drug Delivery

- Market growth of Polymers

## Polymer Architecture

Polymers occupy a major portion of materials used for controlled release formulations and drug-targeting systems because this class of materials presents seemingly endless diversity in topology and chemistry. This is a crucial advantage over other classes of materials to meet the ever-increasing requirements of new designs of drug delivery formulations. The polymer architecture (topology) describes the shape of a single polymer molecule. Every natural, semi natural, and synthetic polymer falls into one of categorized architectures: linear, graft, branched, cross-linked, block, star-shaped, and dendron/dendrimer topology. Although this topic spans a truly broad area in polymer science.

The basic structure of a polymer is a chain. The chain is formed by the backbone of the polymer. It may be decorated with pendant groups, such as methyls in polypropylene, chlorines in polyvinyl chloride or phenyls in polystyrene, but the basic structure is like a long string.

Sometimes, the architecture of a polymer can be more complex than that. For example, low-density polyethylene has branches coming off the main chain. Consequently, it is described as a branched polymer.

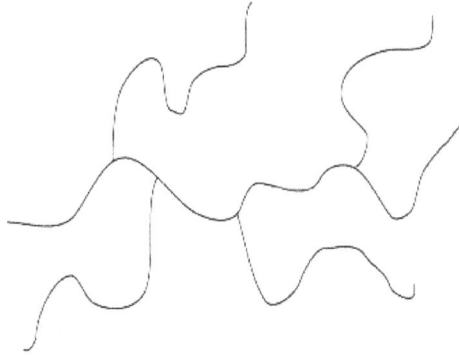

In some cases, branching is so extensive that the polymer does not resemble a chain at all. Some of these highly-branched polymers are called dendrimers. Dendrimers are tree-like structures that branch out in all directions from a central point. Viewed from a distance, a dendrimer would have an overall shape that is more like a fuzzy ball than like a coiled chain.

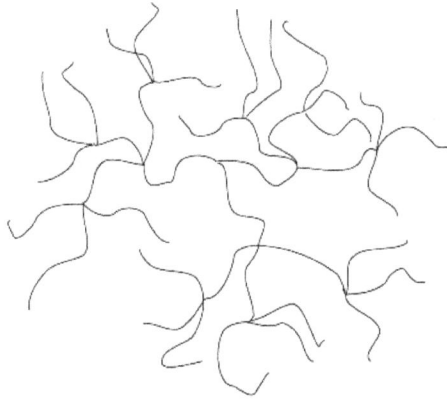

If the branches on one branched chain connect directly to other chains, then the polymer is said to be cross linked. Cross linked polymers can have extra strength and rigidity because they are less flexible than polymers in which the chains are able to move independently. However, just the right amount of crosslinking can make a polymer more elastic, making it bounce back to its original form after it is flexed.

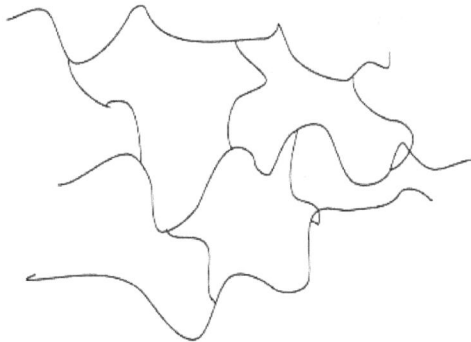

Crosslinking can actually be of two different types: "chemical" or "physical". Chemical crosslinks are permanent covalent bonds between the chains. Physical crosslinks are more temporary; they are formed by strong intermolecular attractions, such as hydrogen bonds, between chains.

## Copolymers

The structure of polymers can vary in how their chains are arranged. Additional structural complexity arises if the polymer is a copolymer. Regular polymers are made from one kind of monomer. Copolymers are formed from two different monomers.

If there are two different monomers, they may be arranged in different ways along the chain. Maybe the two monomers simply alternate, one after the other: that's called an alternating copolymer. Maybe they are distributed at random: that's a random or statistical copolymer. Maybe they are clustered in two separate groups: that's called a block copolymer.

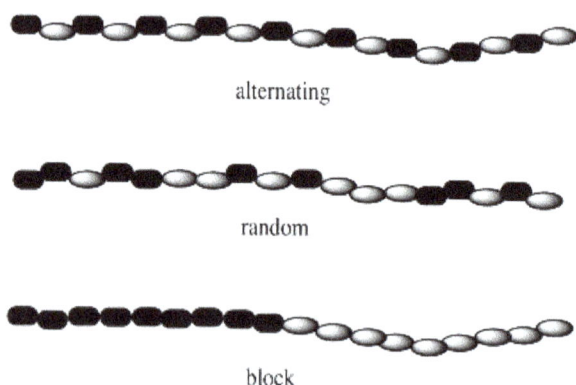

alternating

random

block

These copolymers are different from polymer blends. Polymer blends are just mixtures of two different kinds of polymers. Both polymer blends and copolymers are important. Either one might allow different properties of each polymer, such as strength and flexibility, to be brought together in one material.

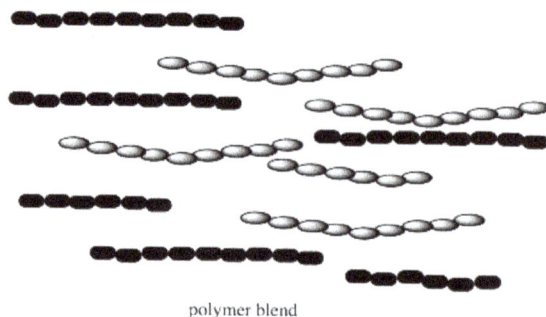

polymer blend

There are different kinds of block copolymers, depending on how the blocks of each monomer are arranged in the chain. The simplest kind of chain would contain one long block of the first monomer followed by one long block of the other. That's called a diblock (or AB diblock) copolymer . If two blocks of the same polymer are found on the ends, with a different one in the middle, that would be called an ABA triblock copolymer. If there were three different blocks from three different monomers, we would have an ABC triblock copolymer instead. There are lots of possible variations.

AB diblock

ABA triblock

ABA tetrablock

Sometimes, the blocks in a block copolymer are not found along the chain. For example, in some cases, the entire backbone may be composed of one monomer, with pendant chains attached to the backbone. A material with this type of architecture is called a graft copolymer, or sometimes a comb or a brush copolymer.

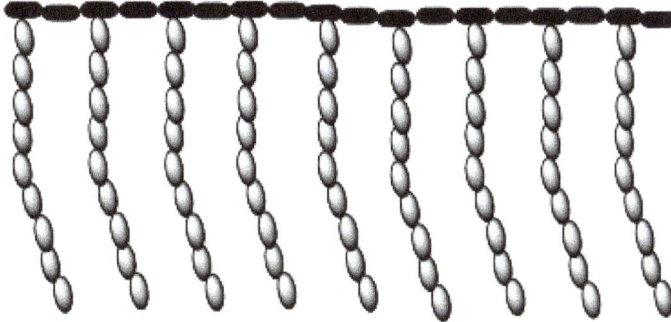

## Effect of Architecture on Physical Properties

In general, the higher degree of branching, the more compact a polymer chain is. Branching also affects chain entanglement, the ability of chains to slide past one another, in turn affecting the bulk physical properties. Long chain branches may increase polymer strength, toughness, and the glass transition temperature ($T_g$) due to an increase in the number of entanglements per chain. A random and short chain length between branches, on the other hand, may reduce polymer strength due to disruption of the chains' ability to interact with each other or crystallize.

An example of the effect of branching on physical properties can be found in polyethylene. High-density polyethylene (HDPE) has a very low degree of branching, is relatively stiff, and is used in applications such as bullet-proof vests. Low-density polyethylene (LDPE), on the other hand, has significant numbers of both long and short branches, is relatively flexible, and is used in applications such as plastic films.

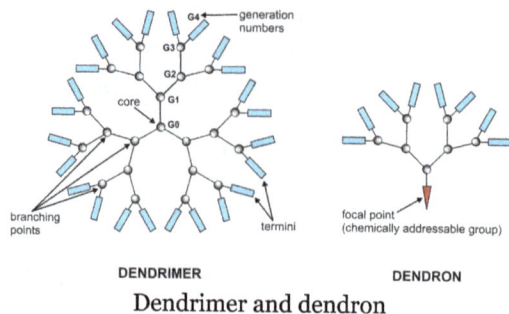

Dendrimer and dendron

Dendrimers are a special case of branched polymer where every monomer unit is also a branch point. This tends to reduce intermolecular chain entanglement and crystallization. A related architecture, the dendritic polymer, are not perfectly branched but share similar properties to dendrimers due to their high degree of branching.

The degree of branching that occurs during polymerisation can be influenced by the functionality of the monomers that are used. For example, in a free radical polymerisation of styrene, addition of divinylbenzene, which has a functionality of 2, will result in the formation of branched polymer.

## Branched Polymers

The properties of polymers are strongly affected by their molecular weight and molecular weight distribution as well as by their chain architecture, particularly by the amount of branching. The effect of branching on the polymer properties depends on the number and length of the branches. Short branches interfere with the formation of crystals, that is, they reduce the amount of crystallinity whereas long branches undergo side chain crystallization because they are able to form lamellar crystals of their own. In the case of polyethylene, significant side chain crystallization can be expected around 40 carbon atoms. Long side chains also have a noticeable effect on the flow properties of the polymer, particularly when the length of the branches exceeds the average critical entanglement length. In that case, even a small amount of branching will greatly affect the processing properties.

Branching plays an important role in the performance of polyolefins. For example, linear polyethylene has a high degree of crystallinity and rather poor mechanical properties. Even a small amount of long-chain branches can significantly improve the mechanical properties and the processability of a polyolefin. This is particularly true when the polyolefin has a narrow molecular weight distribution and a high degree of crystallinity. Both the degree of branching as well as the length of the branches affects the density which can vary considerably. Typically, the higher the density of the polymer the higher the degree of crystallinity and the stiffer, harder, and stronger the polymer.

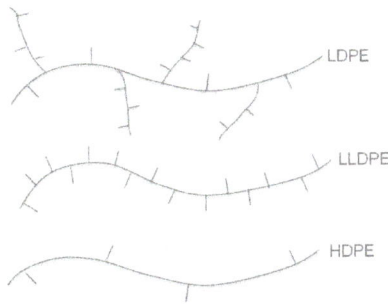

Today, many grades of polyethylene are produced with single-site metallocene catalysts. These polymers are mostly linear and have a high degree of crystallinity. However, these polymers are rather brittle. To increase the flexibility, ethlyene is typically copolymerized with low molecular weight alkenes such as propene, butene-1, hexene-1, 4-methyl-pentene-1 or octene-2 which introduce short chain branches on the mostly linear polymer chain. In general, the toughness and stress crack resistance increases with increasing chain length of the branches. Thus, copolymerization of ethylene with hexene-1 produces tougher polymers than copolymerization with propylene.

The molecular weight distribution and degree and number of chain branching also affect the melt flow properties such as shear thinning behavior and melt flow index. Because of significant differences in molecular structure between highly branched low density polyethylene (LDPE) and linear low and high density polyethylene (LLDPE, HDPE), in particular in molecular weight and chain branching, there are significant differences in the rheological behavior between these types of polyolefins. The effect of long-chain branching on the viscosity was analyzed by Bueche. He found following relationship:

$$\eta_b / \eta_l = g^{3.5}$$

where $\eta_b$ is the viscosity of the branched polymer and $\eta_l$ that of a chemically similar linear polymer of same molecular weight. The parameter $g$ is the ratio of the mean square radii of the branched and linear polymer. Since $g$ is less than unity, Bueche's relation predicts a lower viscosity for long-branched polymers.

## Branching in Radical Polymerization

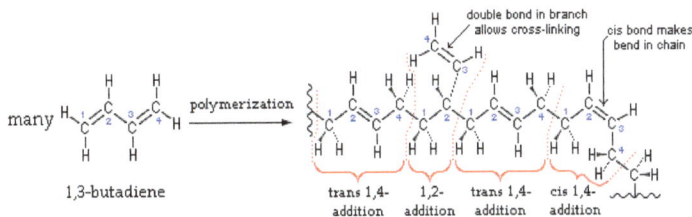

Polymerization of 1,3-butadiene

In free radical polymerization, branching occurs when a chain curls back and bonds to an earlier part of the chain. When this curl breaks, it leaves small chains sprouting from the main carbon backbone. Branched carbon chains cannot line up as close to each other as unbranched chains can. This causes less contact between atoms of different chains, and fewer opportunities for induced or permanent dipoles to occur. A low density results from the chains being further apart. Lower melting points and tensile strengths are evident, because the intermolecular bonds are weaker and require less energy to break.

The problem of branching occurs during propagation, when a chain curls back on itself and breaks - leaving irregular chains sprouting from the main carbon backbone. Branching makes the polymers less dense and results in low tensile strength and melting points. Developed by Karl Zieglerand Giulio Natta in the 1950s, Ziegler-Natta catalysts (triethylaluminium in the presence of a metal(IV) chloride) largely solved this problem. Instead of a free radical reaction, the initial ethene monomer inserts between the aluminium atom and one of the ethyl groups in the catalyst. The polymer is then able to grow out from the aluminium atom and results in almost totally unbranched chains. With the new catalysts, the tacticity of the polypropene chain, the alignment of alkyl groups, was also able to be controlled. Different metal chlorides allowed the selective production of each form i.e., syndiotactic, isotactic and atactic polymer chains could be selectively created.

However, there were further complications to be solved. If the Ziegler-Natta catalyst was poisoned or damaged then the chain stopped growing. Also, Ziegler-Natta monomers have to be small, and it was still impossible to control the molecular mass of the polymer chains. Again new catalysts, the metallocenes, were developed to tackle these problems. Due to their structure they have less premature chain termination and branching.

## Branching Index

The branching index measures the effect of long-chain branches on the size of a macromolecule in solution. It is defined as $g = <sb2>/<sl2>$, where sb is the mean square radius of gyration of the branched macromolecule in a given solvent, and sl is the mean square radius of gyration of an otherwise identical linear macromolecule in the same solvent at the same temperature. A value greater than 1 indicates an increased radius of gyration due to branching.

## Star Polymer

Star polymers consist of several linear polymer chains connected at one point. Star molecules prepared by anionic polymerization had been examined prior to the development of CRP. However due to the limitations of ionic polymerization the composition and functionality of the materials were limited but their compact structure and globular shape provide them with a set of unique properties, such as low solution vis-cosity, and the core shell architecture enabled several potential applications spanning a range from thermoplastic elastomer's to drug carriers.

Based on the chemical compositions of the arm species, star polymers can be classi-fied into two categories: homo-arm (or regular) star polymers or mikto-arm (or het-eroarm) star copolymers. Homo-arm star polymers consist of a symmetrical structure comprising radiating arms with identical chemical composition and similar molecular weight. In contrast, a miktoarm star molecule contains two or more arm species with different chemical compositions and/or molecular weights and/or different peripheral functionality. There are several approaches that can be employed for synthesis of star copolymers.

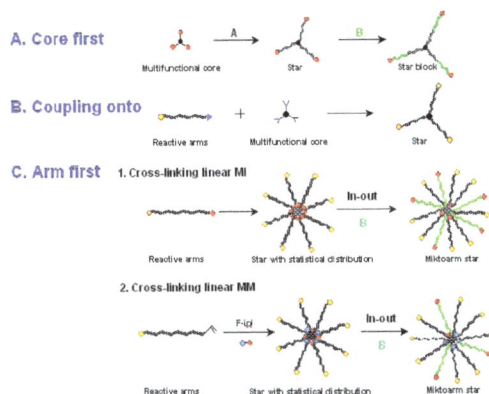

As suggested above, star polymers can be synthesized by variations on one of three methods:

- The *"core-first"* approach, where the controlled polymerization is conducted from either a well defined initiator with a known number of initiating groups or a less well defined multifunctional macromolecule.

- An approach that until recently had not received as much attention is the *"cou-pling onto"* approach where a tele-functional linear molecule is reacted with a preformed core molecule containing complementary functionality. In order to improve the coupling efficiency, a highly efficient organic coupling reaction required, such as click reactions. The preparation of stars using click chemistry could be applied to almost any of the strategies discussed on this page.

- There are two approaches to the *"arm-first"* synthesis of star polymers. One is where a linear "living" copolymer chain, or added macroinitiator, is linked by continuing the copolymerization of the mono-functional macroinitiator with a divinyl monomer forming a cross linked core.

- The other, is the direct copolymerization of a macromonomer with a divinyl monomer in the presence of a low molecular weight initiator.

• A combination of "arm-first" and "core-first" methods is particularly useful for synthesis of miktoarm star copolymers. One employs the retained initiating functionality in the formed "arm first" core to initiate the polymerization of a second monomer in a "grafting out" or a "grafting from" copolymerization.

## Core First

The following section on nano-composites formed by grafting from the surface of a functionalized particle is an evolution of the "core first" approach to synthesis of star macromolecules. Initially a low molecular weight multi-functional molecule was used in a "grafting from" reaction to form star macromolecules with a well defined number of arms. Use of hexakis(chloromethyl) benzene as a well defined multifunctional initiator for the polymerization of styrene provided the first multi-arm polymer prepared using ATRP. The composition of the core and arms were quickly expanded.

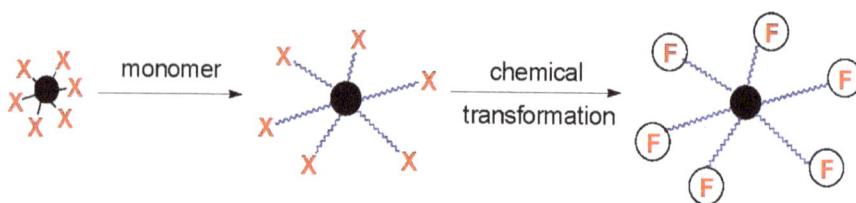

Since the tethered chains in a grafting from reaction retain their terminal functionality they could be chain extended to form star block copolymers and/or the radically transferable atoms on the chain ends (X) could be converted to other functional groups suitable for post-polymerization functionalization reactions.

Less well-defined multi arm star structures were prepared by grafting from a soluble multifunctional hyperbranched core. The core was prepared by polymerizing a molecule containing both a reactive group suitable to initiate a CRP and a polymerizable double bond, an initiator/monomer or inimer. Self-condensing vinyl polymerization using ATRP has been applied to inimers, i.e. monomers that also contain activated halogen atoms, such as α-bromoesters and benzyl halides. For example, when 2-bromo-propionyloxyethyl acrylate was polymerized in the presence of a copper catalyst with 4,4'-dinonyl-2,2'-bipyridine ligands, a hyperbranched polymer with degree of branching ~0.5 and $DP = 78$ was obtained. Each soluble hyperbranched polymer contained on average 78 active bromopropionyloxy groups which were used to initiate the polymerization of methyl acrylate forming a star with 78 arms.

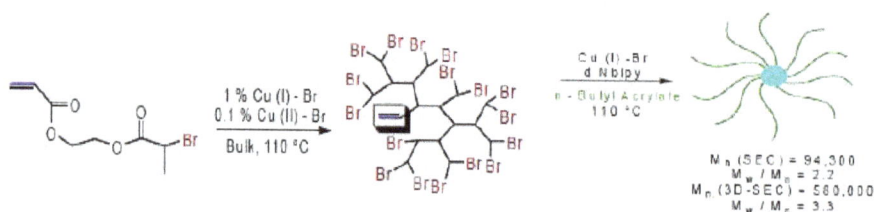

A simple sequential polymerization of a crosslinker followed by polymerization of a monomer provides a broadly applicable approach to star copolymers. This novel method, termed "star from in situ generated core", belongs to the broader category of "core-first" methodology and presents an alternative strategy for star synthesis when compared to the traditional "arm-first" method, in which monomer is polymerized first followed by formation of the core by (co)polymerization of a cross-linker.

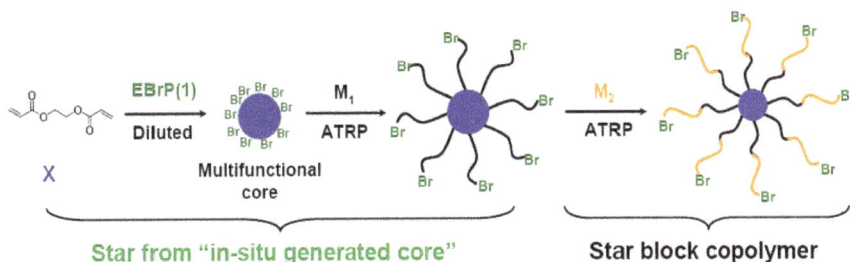

Star from "in-situ generated core"        Star block copolymer

To illustrate this new concept for the synthesis of star polymers using controlled radical polymerization techniques, ATRP was applied to the homopolymerization of a cross linker, ethylene glycol diacrylate (EGDA) to generate a multifunctional cross linked core (nanogel). Monovinyl monomers were added to the reaction mixture at high conversion of the cross linker and were polymerized from the poly EGDA nanogel macro-initiator (MI) to form the star arms. A spectrum of different acrylate monomers were polymerized from the first formed core providing star polymers with different compositions for the arms.

Several parameters affected the structure of the star, such as the initial concentration of the cross linker, the molar ratio of cross linker to initiator, and the moment of addition of the monovinyl monomer to the ongoing polymerization reaction. The star polymers preserved the initiating sites at the chain ends, and they were further used as star MIs for arm extension by polymerization of a second monovinyl monomer to form a star block copolymer.

Indeed this approach of preparing well defined polymers by copolymerization of a cross linker and a monomer can be expanded to provide access to a full range of stars/gels/networks.

## Arm First

## Macroinitiator Approach

The "arm first" approach forms the core of the star macromolecule by coupling monofunctional "living" polymeric chains with a difunctional reagent and was first applied to living anionic polymerization. A similar approach was also successful using ATRP. Initially the simple chain extension of a linear macroinitiator with a crosslinker provided star macromolecules with broad polydispersity as a result of star-star coupling reactions.

There are several parameters in an ATRP that should be carefully controlled in order to maximize the yield of stars and prevent/reduce star-star coupling reactions. Some detailed studies have been carried out on the coupling of monofunctional polystyrenes and polyacrylates with divinylbenzene (DVB) and di(meth)acrylates to prepare star polymers and the following guidelines were developed:

- The ratio of di-functional reagent to growing chains seems to be optimal in the range of 10-20.

- Monomer conversion (or reaction time) has to be carefully controlled and stopped before star-star coupling occurs. It seems that ~5% of arms can not be incorporated into the star macromolecules under typical one step conditions.

- Higher yields of stars are observed for polyacrylates than for polystyrenes. This may be attributed to a higher proportion of terminated chains (due to slower propagation and higher concentration of radicals) in styrene polymerization under "standard" ATRP conditions. (The use of ARGET ATRP for the synthesis of the active growing polymer chains should change both these observations.)

- The choice of the diffunctional reagent is important and reactivity should be similar to, or lower than that of the arm-building monomers.

- Halogen exchange slightly improves the efficiency of star formation.

- Solvent, temperature and catalyst concentration should be also optimized.

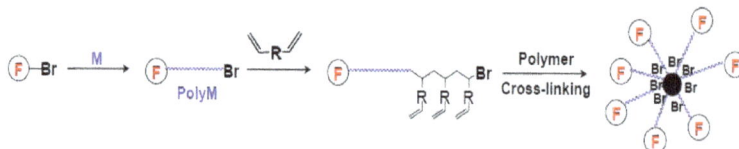

Under typical conditions, ~50 arms are coupled into star structures. Apparent molecular weights, measured by GPC, show a 10-20-fold increase of molecular weight (from $M_n = 10,000$ to $M_n \sim 150,000$) but light scattering indicates that the stars have a much higher molecular weight $(M_n \sim 500,000)$.

A range of initiators, monomers and cross linkers was examined for the preparation of star molecules with peripheral functionality and cores of differing phobicity.

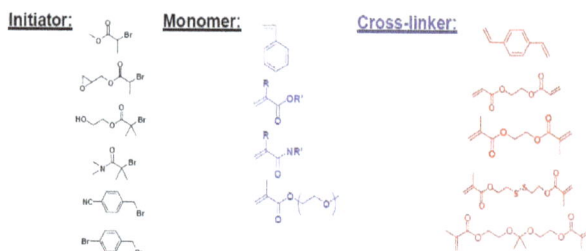

Another approach to form star polymers in high yield was via cross-linking self organized amphiphilic macroinitiators by AGET ATRP in aqueous dispersed media. Linear poly(ethylene oxide)-b-polystyrene (PEO-PS-Cl) block copolymers were the arm precursors. The amphiphilic block copolymers with halogen chain-end functionality formed divinyl cross-linker swollen micelles in water and then were cross-linked by the cross-linkers when polymerization was initiated. Due to the formation of micelles before the polymerization was initiated, star-star or star-linear chain reactions were not required for the star formation. This suppressed star-star coupling reactions and resulted in the formation of star polymers with low PDI (Mw/Mn < 1.1) but still high molecular weight (over 1000 kg/mol).

## Macromonomer Approach

Another approach to arm first star copolymers, with available core containing initiating functionality, is the CRP of higher molecular weight macromonomers in a pure homo-polymerization initiated with a small molecule initiator which usually leads to a "brush molecule" with a degree of polymerization of 10-25, which from a topological standpoint can be considered a star.

An alternate approach is a copolymerization of macromonomers with a divinyl cross-linking monomer.

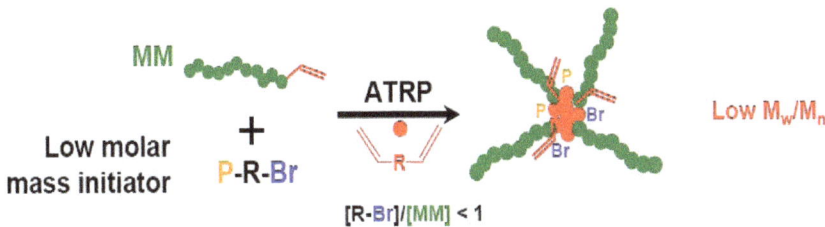

The sequential addition of additional initiator and crosslinker to the reaction increases the number of macromonomer units incorporated into each star.

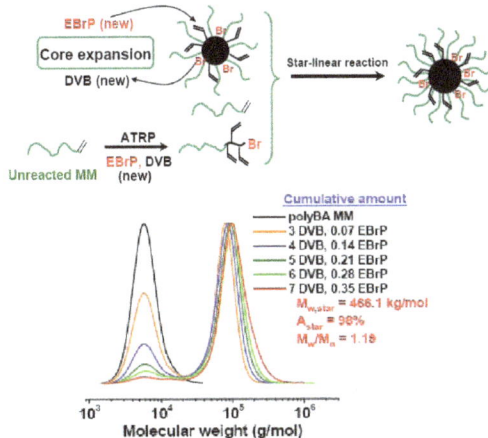

A potential application for homo-arm stars was illustrated by preparation of two well defined star macromolecules with two oppositely charged arm structures; a poly[2-(dimethylamino)ethyl methacrylate] (PDMAEMA) star and a poly(acrylic acid) (PAA) star with cross-linked cores. Exploitation of the electrostatic interactions between the polyelectrolyte arms of PDMAEMA star and PAA star polymers generated alternating polyelectrolyte multilayer LbL films using layer-by-layer (LbL) assembly.

## Mikto-arm Star Copolymers

## In-out Method

Both approaches to form star molecules using the arm first approach retain the initiating functionality within the core of the star and therefore provide a simple approach to form mikto-arm stars by conducting a controlled polymerization of a second monomer from the accessible initiator functionality in the core.

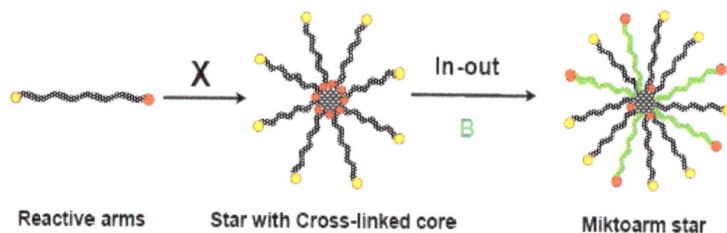

Cross-linking linear MI    (Macroinitiator + X)

Reactive arms      Star with Cross-linked core      Miktoarm star

As noted above one arm-first procedure involves the synthesis of a linear mono-functional polymer chain polyA, which is used as a macroinitiator (MI) in a subsequent cross-linking reaction using a divinyl compound to produce a $(polyA)_n$-polyX star polymer, where polyX represents the core of the star polymer and n is the number of polyA arms. The initiating sites are preserved within the core of the star polymer (i.e., alkyl halide groups in ATRP) and the $(polyA)_n$-polyX star polymer can be used as a multifunctional star initiator in a chain extension reaction with a different monomer, B, to yield a miktoarm star copolymer, $(polyA)_n$-polyX-$(polyB)_m$. This combination method for synthesizing miktoarm star copolymers was termed the "in-out" method.

The efficiency of initiation of the second set of arms is dependent on the compactness of the first formed core with less densely crosslinked cores providing more efficient initiation i.e., a greater fraction of the encapsulated initiator sites, for the grafting out polymerization.

Normally, one seeks to, or has to, form a chemically stable core. However, it is possible to select a crosslinking agent with a degradable link between the two functional crosslinking groups and prepare a material with a degradable core.

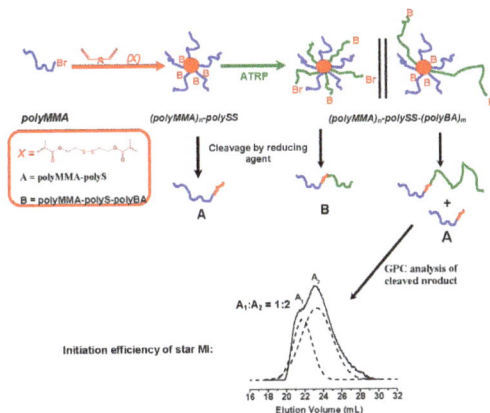

polyMMA    (polyMMA)$_n$-polySS    (polyMMA)$_n$-polySS-(polyBA)$_m$

$X =$

A = polyMMA-polyS
B = polyMMA-polyS-polyBA

Cleavage by reducing agent

GPC analysis of cleaved product

$A_1$:$A_2$ = 1:2

Initiation efficiency of star MI:

16   18   20   22   24   26   28   30   32
Elution Volume (mL)

This was accomplished by linking the first formed arms with a dimethacrylate crosslinker containing a disulfide link between the methacrylate units. The mikto-arm star copolymer could be degraded in a reducing environment to form a mixture of an AB block copolymer and some residual A-homopolymer. The ratio between the block copolymer and homopolymer gave a direct measurement of the initiating efficiency of the constrained core initiating units; in this example it was only ~20%.

Miktoarm star synthesis: adding cross-linker after formation of several MIs

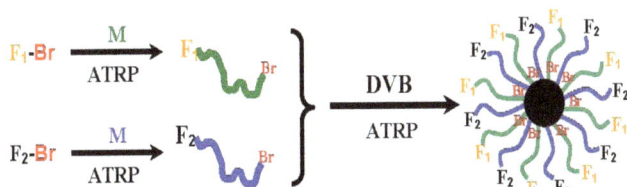

$F_1$-Br $\xrightarrow[\text{ATRP}]{M}$    $F_2$-Br $\xrightarrow[\text{ATRP}]{M}$    $\xrightarrow[\text{ATRP}]{DVB}$

Liquid chromatography under the critical conditions (LCCC) for each of the homopolymers arms provided proof that the star structure a miktoarm star copolymer containing two or more compositionally different arm species in one molecule, and was not a mixture of different homo-arm star polymers. Miktoarm star copolymers containing five kinds of arms were synthesized for the first time by copolymerizing a mixture of five linear MIs with different chemical composition. When linear MMs were partially or completely used as arm precursors instead of MIs, miktoarm star copolymers with high star yield and low polydispersity were successfully synthesized.

A recent example of a combination of an "arm first" and a "core first" approach is "grafting onto" a functionalized core via "click" chemistry. Three-arm and four-arm star polystyrene polymers were synthesized by a combination of ATRP and click coupling chemistry. The click reaction between an azido-terminated polystyrene (PS-N3) and an alkyne-containing multifunctional compound proved to be fast and efficient. When an azido-terminated polystyrene polymer was reacted with a trialkyne-containing or tetraalkyne-containing compound, the yields of 3-arm star and 4-arm star polymers were around 90% and 83%, respectively.

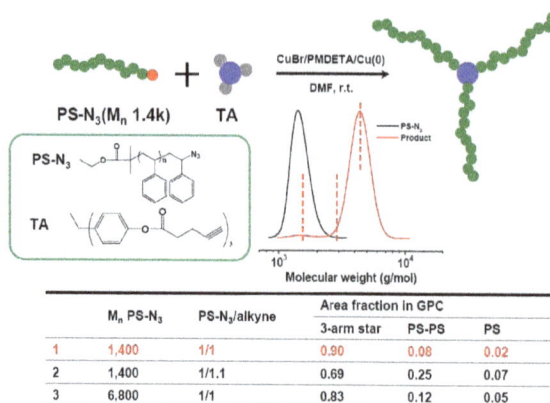

| | $M_n$ PS-N$_3$ | PS-N$_3$/alkyne | Area fraction in GPC | | |
|---|---|---|---|---|---|
| | | | 3-arm star | PS-PS | PS |
| 1 | 1,400 | 1/1 | 0.90 | 0.08 | 0.02 |
| 2 | 1,400 | 1/1.1 | 0.69 | 0.25 | 0.07 |
| 3 | 6,800 | 1/1 | 0.83 | 0.12 | 0.05 |

The influence of several parameters on the efficiency of the click coupling reaction was studied, including the molecular weight of the PS-N3 polymer, the presence of an added reducing agent, Cu°, and the stoichiometry between the azido and alkynyl groups. The results indicated that the yield of the coupled product was higher when a lower molecular weight PS-N3 was employed in conjunction with a small amount of reducing agent, and the molar ratio of azido and alkynyl groups was close to 1.

## Graf Polymer

Graft polymers are segmented copolymers with a linear backbone of one composite and randomly distributed branches of another composite. The picture labeled "graft polymer" shows how grafted chains of species B are covalently bonded to polymer species A. Although the side chains are structurally distinct from the main chain, the individual grafted chains may be homopolymers or copolymers. Graft polymers have been synthesized for many decades and are especially used as impact resistant materials, thermoplastic elastomers, compatibilizers, or emulsifiers for the preparation of stable blends or alloys. One of the more well known examples of a graft polymer is high impact polystyrene, which consists of a polystyrene backbone with polybutadiene grafted chains.

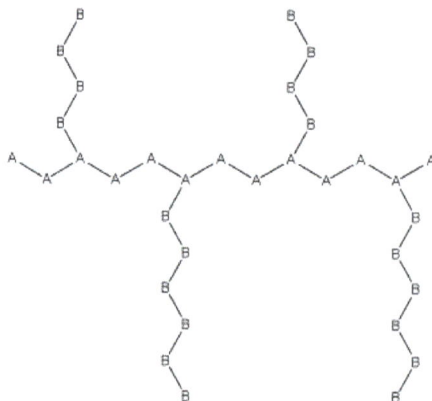

The graft copolymer consists of a main polymer chain or backbone
(A) covalently bonded to one or more side chains (B)

## General Properties

Graft copolymers are a branched copolymer where the components of the side chain are structurally different than that of the main chain. Graft copolymers containing a larger quantity of side chains are capable of wormlike conformation, compact molecular dimension, and notable chain end effects due to their confined and tight fit structures. The preparation of graft copolymers has been around for decades. All synthesis methods can be employed to create general physical properties of graft copolymers. They can be used for materials that are impact resistant, and are often used as thermoplastics elastomers, compatibilizers or emulsifiers for the preparation of stable blends or alloys. Generally, grafting methods for copolymer synthesis results in materials that are more thermostable than their homopolymer counterparts. There are three methods of synthesis, grafting to, grafting from, and grafting through, that are used to construct a graft polymer.

## Synthesis Methods

There are many different approaches to synthesizing graft copolymers. Usually they employ familiar polymerization techniques that are commonly used such as atom transfer radical polymerization (ATRP), ring-opening metathesis polymerization (ROMP), anionic and cationic polymerizations, and free radical living polymerization. Some other less common polymerization include radiation-induced polymerization, ring-opening olefin metathesis polymerization, polycondensation reactions, and iniferter-induced polymerization.

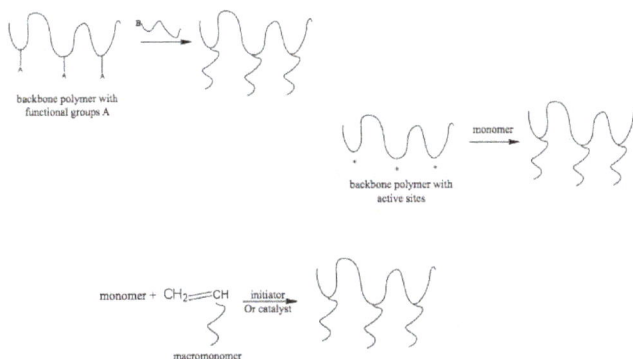

The three common methods of synthesis: grafting to (top left), grafting from (middle right), grafting through (bottom left), and their generalized reaction scheme are featured.

## Grafting to

The grafting to method involves the use of a backbone chain with functional groups A that are distributed randomly along the chain. The formation of the graft copolymer originates from the coupling reaction between the functional backbone and the end-groups of the branches that are reactive. These coupling reactions are made possible by modifying the backbone chemically. Common reaction mechanisms used to synthesize these copolymers include free-radical polymerization, anionic polymerization, atom-transfer radical-polymerization, and living polymerization techniques.

Copolymers that are prepared with the grafting-to method often utilize anionic polymerization techniques. This method uses a coupling reaction of the electrophilic groups of the backbone polymer and the propagation site of an anionic living polymer. This method would not be possible without the generation of a backbone polymer that has reactive groups. This method has become more popular with the rise of click chemistry. A high yield chemical reaction called atom transfer nitroxide radical coupling chemistry is for the grafting-to method for polymerization.

## Grafting from

In the grafting-from method, the macromolecular backbone is chemically modified in order to introduce active sites capable of initiating functionality. The initiating sites can be incorporated by copolymerization, can be incorporated in a post-polymerization reaction, or can already be a part of the polymer. If the number of active sites along the backbone participates in the formation of one branch, then the number of chains grafted to the macromolecule can be controlled by the number of active sites. Even though the number of grafted chains can be controlled, there may be a difference in the lengths of each grafted chain due to kinetic and steric hindrance effects.

Grafting from reactions have been conducted from polyethylene, polyvinylchloride, and polyisobutylene. Different techniques such as anionic grafting, cationic grafting, atom-transfer radical polymerization, and free-radical polymerization have been used in the synthesis of grafting from copolymers.

Graft copolymers that are employed with the grafting-from method are often synthesized with ATRP reactions and anionic and cationic grafting techniques.

## Grafting Through

The grafting through, also known as the macromonomer method, is one of the simpler ways of synthesizing a graft polymer with well defined side chains. Typically a monomer of a lower molecular weight is copolymerized with free radicals with an acrylate functionalized macromonomer. The ratio of monomer to macromonomer molar concentrations as well as their copolymerization behavior determines the number of chains that are grafted. As the reaction proceeds, the concentrations of monomer to macromonomer change causing random placement of branches and formation of graft copolymers with different number of branches. This method allows for branches to be added heterogeneously or homogeneously based on the reactivity ratio of the terminal functional group on the macromolecular to the monomer. The difference in distribution of grafts has significant effects on the physical properties of the grafted copolymer. Polyethylene, polysiloxanes and poly(ethylene oxide) are all macromonomers that have been incorporated in a polystyrene or poly(methyl acrylate) backbone.

The macromonomer (grafting through) method can be employed using any known polymerization technique. Living polymerizations give special control over the molecular weight, molecular weight distribution, and chain-end functionalization.

## Applications

Graft copolymers became widely studied due to their increased number of applications due to their unique structures relative to other copolymers such as alternating, periodic, statistical, and block copolymers which generally linear chains.

Some common applications of graft copolymers include:

- Membranes for the separation of gases or liquids

- Hydrogels

- Drug deliverers

- Thermoplastic elastomers

- Compatibilizers for polymer blends

- Polymeric emulsifiers

- Impact resistant plastics

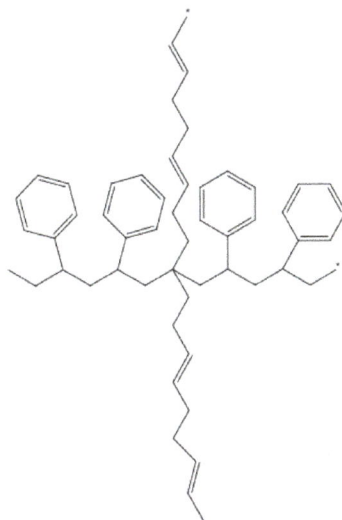

High Impact Polystyrene (HIPS) consists of the polystyrene backbone with polybutadiene chains branching from it in each direction.

## High Impact Polystyrene

CD case made from general purpose polystyrene (GPPS) and high impact
polystyrene in the black portion (HIPS)

High impact polystyrene (HIPS) was discovered by Charles F. Fryling in 1961. HIPS is a low cost, plastic material that is easy to fabricate and often used for low strength structural applications when impact resistance, machinability, and low cost are required. Its major applications include machined prototypes, low-strength structural components, housings, and covers. In order to produce the graft polymer, polybutadiene (rubber) or any similar elastomeric polymer is dissolved in styrene and polymerized. This reaction allows for two simultaneous polymerizations, that of styrene to polystyrene and that of the graft polymerization of styrene-rubber. During commercial use, it can be prepared by graft copolymerization with additional polymer to give the product specific characteristics. The advantages of HIPS includes:

- FDA compliant

- Good impact resistance

- Excellent machinability

- Good dimensional stability

- Easy to paint and glue

- Low cost

- Excellent aesthetic qualities

## New properties as a Result of Grafting

By grafting polymers onto polymer backbones, the final grafted copolymers gain new properties from their parent polymers. Specifically, cellulose graft copolymers have various different applications that are dependent on the structure of the polymer grafted onto the cellulose. Some of the new properties that cellulose gains from different monomers grafted onto it include:

- Absorption of water

- Improved elasticity

- Hydrophilic/Hydrophobic character

- Ion-exchange

- Dye adsorption capabilities

- Heat Resistance

- Thermosensitivity

- pH sensitivity

- Antibacterial effect

These properties give new application to the ungrafted cellulose polymers that include:

- Medical body fluid absorbent materials.

- Enhanced moisture absorbing ability in fabrics.

- Permselective membranes.

- Stronger nucleating properties than ungrafted cellulose, and adsorption of hazardous contaminants like heavy metal ions or dyes from aqueous solutions by temperature swing adsorption.

- Sensors and optical materials.

- Reducing agents for various carbonyl compounds.

## Comb Polymer

Comb polymer, which consists of a hydrophobic poly(methyl methacrylate) (PMMA) backbone with hydrophilic hydroxy-poly(ethylene oxide) (HPOEM) side chains, is a tool that has many possible applications for the study of liver cell adhesion and signaling. This polymer has the unique properties of being cell resistant and chemically versatile such that various cell ligands can be coupled to its side chains. These properties allow adhesion through specific cell receptors to be studied without the effect of background adhesion to adsorbed proteins. By taking advantage of the ability to target specific receptors the comb polymer could be used as a powerful sorting tool. Sorting could be accomplished by finding cell type specific adhesion ligands. Several possible such ligands were screened. A ligand containing the tripeptide sequence RGD was found to elicit a strong cell adhesion response. However, this ligand is adherent to many cell types of the liver and would not be suitable for sorting purposes. Other cell type specific ligands tested showed little to no affinity for liver cell adhesion. Additionally, the comb was utilized to study $\alpha_5\beta_1$ integrin-specific hepatocyte adhesion and the effect of Epidermal Growth Factor on adhesion. $\alpha_5\beta_1$ integrin adhesion was mediated using a novel branched peptide, SynKRGD. This peptide consists of a linear peptide sequence containing RGDSP and the synergy site sequence PHSRN connected by the sequence GGKGGG. By utilizing the amine side group of Lysine a GGC branch was added. The terminal cysteine was used to conjugate SynKRGD to comb polymer surfaces using N-(p-Maleimidophenyl) isocyanate (PMPI) chemistry. EGF has a great potential to benefit the field of tissue engineering due to its influence on cell(cont.) proliferation, migration, and differentiation. EGF is also known to have a de-adhesive effect in some cell types. Hepatocytes were studied on comb surfaces of variable SynKRGD densities with and without the presence of EGF in the media. Distinct morphological differences were observed for hepatocytes on substrates of varying adhesivity with and without the presence of EGF. EGF was found to have a de-adhesive effect on $\alpha_5\beta_1$ integrin adhesion in hepatocytes. This effect became more pronounced as substrate adhesiveness increased.

## Brush Polymer

A brush polymer is the name given to a surface coating consisting of polymers tethered to a surface. The brush may be either in a solvated state, where the tethered polymer layer consists of polymer and solvent, or in a melt state, where the tethered chains completely fill up the space available. These polymer layers can be tethered to flat substrates such as silicon wafers, or highly curved substrates such as nanoparticles. Also, polymers can be tethered in high density to another single polymer chain, although this arrangement is normally named a bottle brush.Additionally, there is a separate class of polyelectrolyte brushes, when the polymer chains themselves carry an electrostatic charge.

The brushes are often characterized by the high density of grafted chains. The limited space then leads to a strong extension of the chains. Brushes can be used to stabilize colloids, reduce friction between surfaces, and to provide lubrication in artificial joints.

Polymer brushes have been modeled with Monte Carlo methods, Brownian dynamics simulations, and molecular theories.

## Structure

Polymer molecules within a brush are stretched away from the attachment surface as a result of the fact that they repel each other (steric repulsion or osmotic pressure). More precisely, they are more elongated near the attachment point and unstretched at the free end, as depicted on the drawing.

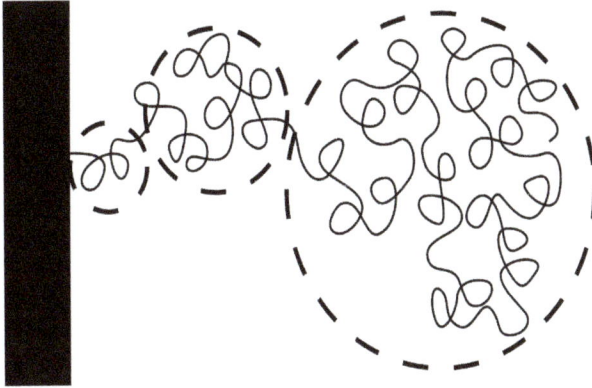

Polymer molecule within a brush. The drawing shows the chain elongation decreasing from the attachment point and vanishing at free end.

The "blobs", schematized as circles, represent the (local) length scale at which the statistics of the chain change from a 3D random walk (at smaller length scales) to a 2D in-plane random walk and a 1D normal directed walk (at larger length scales).

More precisely, within the approximation derived by Milner, Witten, Cates, the average density of all monomers in a given chain is always the same up to a prefactor:

$$\phi(z,\rho) = \frac{\partial n}{\partial z}$$

$$n(z,\rho) = \frac{2N}{\pi}\arcsin\left(\frac{z}{\rho}\right)$$

where $\rho$ is the altitude of the end monomer and $N$ the number of monomers per chain.

The averaged density profile $\epsilon(\rho)$ of the end monomers of all attached chains, convoluted with the above density profile for one chain, determines the density profile of the brush as a whole:

$$\phi(z) = \int_\rho^\infty \frac{\partial n(z,\rho)}{\partial z}\epsilon(\rho)\mathrm{d}\rho.$$

A dry brush has a uniform monomer density up to some altitude $H$. One can show that the corresponding end monomer density profile is given by:

$$\epsilon_{\mathrm{dry}}(\rho,H) = \frac{\rho/H}{Na\sqrt{1-\rho^2/H^2}}$$

where $a$ is the monomer size.

The above monomer density profile $n(z,\rho)$ for one single chain minimizes the total elastic energy of the brush,

$$U = \int_0^\infty \epsilon(\rho)\mathrm{d}\rho \int_0^N \mathrm{d}n \frac{kT}{2Na^2}\left(\frac{\partial z(n,\rho)}{\partial n}\right)^2.$$

regardless of the end monomer density profile $\epsilon(\rho)$.

## From a Dry Brush to any Brush

As a consequence, the structure of any brush can be derived from the brush density profile $\phi(z)$. Indeed, the free end distribution is simply a convolution of the density profile with the free end distribution of a dry brush:

$$\epsilon(\rho) = \int_\rho^\infty -\frac{\mathrm{d}\phi(H)}{\mathrm{d}H}\epsilon_{\mathrm{dry}}(\rho, H)..$$

Correspondingly, the brush elastic free energy is given by:

$$\frac{F_{\mathrm{el}}}{kT} = \frac{\pi^2}{24N^2a^5}\int_0^\infty \left\{-z^3\frac{\mathrm{d}\phi(z)}{\mathrm{d}z}\right\}\mathrm{d}z.$$

This method has been used to derive wetting properties of polymer melts on polymer brushes of the same species and to understand fine interpenetration asymmetries between copolymer lamellae that may yield very unusual non-centrosymmetric lamellar structures.

## Dendronized Polymer

Dendronized polymers are formally a sub-group of comb polymers, where the linear chains of the grafts have been replaced by dendrons as depicted in Figure. The properties of these kinds of polymers are believed to depend on a number of factors, such as type of dendron, attachment density along the backbone, end-group functionality, and degree of polymerization. Because of the large number of properties that can be tailored independently in this type of architecture, several novel applications have been proposed, for example building blocks for nanometer sized construction, efficient light emitting materials, isolating conducting polymers, complexation agents for DNA, degradable drug carriers, and as support for catalysts. If spatially demanding dendrons are attached to a polymeric backbone, the polymer can be forced from a random-coil conformation to a stretched cylindrical.44,46 The multifunctional surface of these dendronized cylinders can be addressed and because of the potentially high bending modulus of these materials, dendronized polymers can be envisioned as molecular reinforcements in, for example, coating applications.

# Synthesis of Dendronized Polymers

The first synthetic route to dendronized polymers was described in 1987 in a patent by Tomalia et al. and since then considerable synthetic effort has been devoted to developing new and efficient routs for these novel materials. Figure displays the two main routes to attaining dendronized polymers.

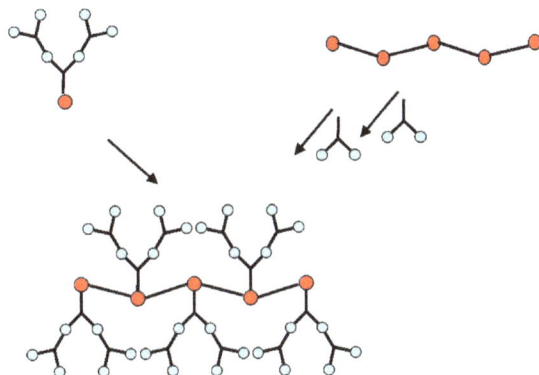

Figure: The macromonomer route (left), and the graft-onto route (right) to dendronized polymers.

In the macromonomer route, dendrons are linked to monomers that contain a polymerizable group either by a convergent or divergent methodology. The macromonomers are then polymerized, forming a dendronized polymer. methods have been employed in the macromonomer route for example radical polymerization, reversible addition-fragmentation chain transfer (RAFT) polymerization, ring-opening metathesis polymerization (ROMP), ring-opening polymerization (ROP), Suzuki polycondensation, insertion polymerization, Stille coupling, Heck coupling, polycondensation, and atom transfer radical polymerisation (ATRP).

The drawback of the macromonomer route is that larger dendrons may shield the polymerizable group and thereby inhibit the polymerization, resulting in either low molecular weight polymers or no reaction at all. Studies have addressed this problem, and propose that by increasing the distance between dendron and polymerizable group, the availability of the reactive group will increase. However, recent studies indicate that the problem of inhibition may be related only to the concentration of the monomer, if the concentration of monomer is kept high enough, high degrees of polymerization can be reached.

In the 'graft-onto' route preformed dendrons are attached to a polymeric backbone. Both convergent and divergent approaches have been employed. The 'graft-onto' route may suffer the drawback of incomplete coupling steps due to steric crowding of the dendrons. To circumvent this problem, large excesses of coupling agent and long reaction times must be employed. The large excesses may lead to purification problems of the product. Divergent approaches to dendronized polymers utilizing either benzylidene-2,2- bis(methoxy) propanoic anhydride or acetonide-2,2-bis(methoxy) propanoic anhydride allow for purification of large excesses of coupling agent by simple extractions. The benefit of combining the 'graft-onto' route with a controlled polymerization technique is that narrowly distributed dendronized polymers emanating from the same backbone can be produced with ease.

## Structure and Applications

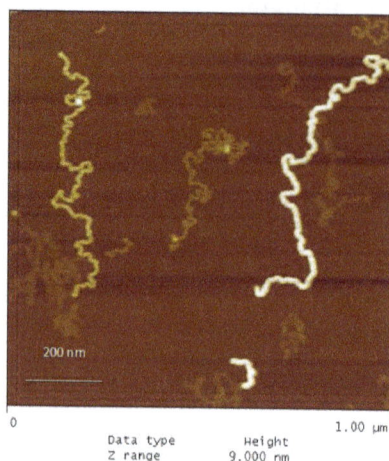

Figure: Atomic force microscopy height image of co-prepared dendronized polymers of generation one through four (PG1-PG4) reflecting the different thicknesses and apparent persistence lengths for each generation

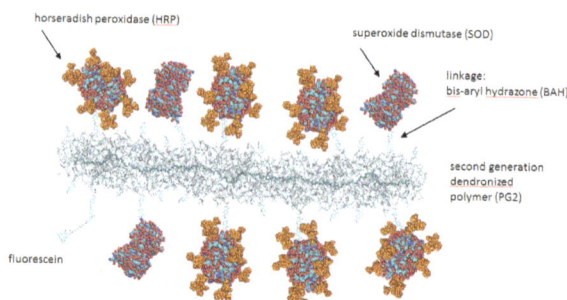

Figure: Schematic representation of a molecular hybrid structure (conjugate) between a dendronized polymer and the two different enzymes HRP (horseradish peroxidase) and SOD (Cu,Zn-superoxide dismutase). PDB (SOD): 1SXA; PDB (HRP): 1ATJ; HRP sugar modification from Gray & Montgomery, *Carbohydrate*.

Dendronized polymers can contain several thousands of dendrons in one macromolecule and have a stretched out, anisotropic structure. In this regard they differ from the more or less spherically shaped dendrimers, where a few dendrons are attached to a small, dot-like core resulting in an isotropic structure. Depending on dendron generation, the polymers differ in thickness as the atomic force microscopy image shows Figure. Neutral and charged dendronized polymers are highly soluble in organic solvents and in water, respectively. This is due to their low tendency to entangle. Dendronized polymers have been synthesized with, e.g., polymethylmethacrylate, polystyrene, polyacetylene, polyphenylene, polythiophene, polyfluorene, poly(phenylene vinylene), poly(phenylene acetylene), polysiloxane, polyoxanorbornene, poly(ethylene imine) (PEI) backbones. Molar masses up to 200 Mio g/mol have been obtained. Dendronized polymers have been investigated for/as bulk structure control, responsivity to external stimuli, single molecule chemistry, templates for nanoparticle formation, catalysis, electro-optical devices, and bio-related applications. Particularly attractive is the use of water-soluble dendronized polymers for the immobilization of enzymes on solid surfaces (inside glass tubes or microfluidic devices) and for the preparation of dendronized polymer-enzyme conjugates.

## Ladder Polymers

A ladder polymer is defined as a "uninterrupted series of rings connected by sterically restrictive links around which rotation cannot occur without the breaking of a bond" .

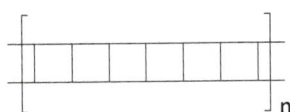

"Classical" ladder Polymer

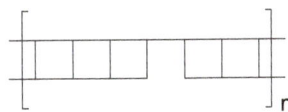

"Imperfect" ladder or step ladder polymer

Conjugated ladder polymers possess a planarized π-electron system which ensures optimum electron delocalization. Theory predicts that such materials will exhibit outstanding electronic properties including siginificant third-order susceptibilities. Ladder polymers are predicted to show a high thermal stability and a high resistance to chemical degradation. In order to have all these unique properties perfect ladder polymers are required. However, in practice, it is often difficult to bring nearly perfect ladder structures for experimental realization. There are many reasons for these difficulties. First, many ladder polymers are insoluble in common solvents and precipitate from solutions before reactions are complete. This deficiency not only leads to the formation of low molecular-weight products but also leads to incompletely cyclized structures. Without a doubt, the synthesis of a perfect ladder polymer is a real synthetic challenge. In the case of ladder polymers, two independent strands have to be generated without defects or crosslinking. In the early 1970s poly(benzimidazobenzophenanthroline) (BBL 17) was synthesized and it was discovered that this ladder polymer is readily processible into durable films using casting or spraying techniques.

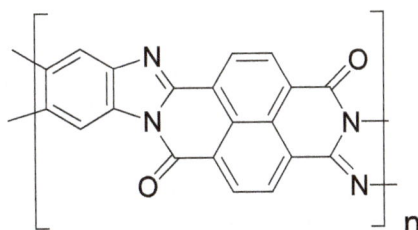

BBL 17

Several other ladder polymers were than synthesized and a few ladder polymers gained some industrial importance as materials for high-temperature applications such as heat-resistant coatings e.g. in aerospace vehicles.

## Synthesis of Ladder Polymers

Attempts to generate ladder polymers have a long history, but ladder polymers prepared at the initial stage showed a poor solubility, due to their rigid structure. Many synthetic efforts have been made to bring soluble ladder polymer to an experimental realization. A new generation of substituted, soluble ladder polymer have been made using the following improved approaches:

- Novel polycondensation methods e.g. repetitive Diels-alder reaction.

- Polymer-analogous cyclization of single-stranded linear prepolymers.

Most of the known synthesis of ladder polymers fall into the these two methods. But it is essential to prevent side-reactions and to take care of an almost quantitative conversion of the starting materials. It is beyond the scope of this chapter to cover all the history of ladder polymers. We will deal with the conjugated poly(para-phenylene)-type (LPPP) ladder polymers. The polymer-analogous cyclization of suitable single-stranded precursors have been successfully applied to synthesize such polyphenylene-based conjugated ladder polymers.

# Two-dimensional Polymers

Two-dimensional polymers are sheet-like, covalently bonded molecular tilings that extend in exactly two dimensions. Unlike 1D polymers, over which chemists can often exert high levels of control during their synthesis, the design and production of 2D polymers presents far greater challenges. Graphene is perhaps the archetypal 2D material, and although it has enjoyed a great deal of attention, the idea of precisely tailoring its size, shape and functionality at the molecular level seems somehow incompatible with something that can be made in a kitchen blender. Perusing the recent literature, you could be forgiven for thinking that graphene was the only 2D material in existence, but other natural and synthetic layered substances have also been exfoliated into 2D sheets.

The unique optical and electronic properties that arise from the 2D structure of graphene have attracted the attention of many researchers, including those interested in modulating its behaviour through chemical functionalization — although there can be some uncertainty regarding the exact nature and placement of functional groups. Bottom-up approaches to functional graphene fragments5 (or so-called nanographenes) look to overcome the problems associated with top-down methods, but are currently limited in terms of the size of sheets that can be formed, and the functionality that can be introduced. Two-dimensional synthetic polymers arguably have the potential to overcome these limitations, particularly if large crystals can be grown.

# References

- Campbell, Neil A.; Brad Williamson; Robin J. Heyden (2006). Biology: Exploring Life. Boston, Massachusetts: Pearson Prentice Hall. ISBN 0-13-250882-6

- Star-copolymers, polymers-with-Specific-Architecture: cmu.edu, Retrieved 25 April 2018

- Feng, Chun; Li, Yongjun; Yang, Dong; Hu, Jianhua; Zhang, Xiaohuan; Huang, Xiaoyu (2011). "Well-defined graft copolymers: from controlled synthesis to multipurpose applications". Chemical Society Reviews. 40 (3): 1282–95. doi:10.1039/b921358a. PMID 21107479

- Polymer-sciences-journals-articles-ppts-list: omicsonline.org, Retrieved 17 July 2018

- Ito, Koichi; Hiroyuki Tsuchida; Akio Hayashi; Toshiaki Kitano (1985). "Reactivity of Poly(ethylene oxide) Macromonomers in Radical Copolymerization". Polymer Journal. 17(7): 827–839. doi:10.1295/polymj.17.827

- Volker Abetz ... [et (2005). Encyclopedia of polymer science and technology (Wird aktualisiert. ed.). [Hoboken, N.J.]: Wiley-Interscience. ISBN 9780471440260

- Future-scope-of-polymer-science: polymerchemistry.chemistryconferences.org, Retrieved 27 May 2018

- Pearce, Eli M. (May 1987). "New commercial polymers 2, by Hans-George Elias and Friedrich Vohwinkel, Gordon and Breach, New York, 1986, 508 pp. Price: $90.00". Journal of Polymer Science Part C: Polymer Letters. 25 (5): 233–234. doi:10.1002/pol.1987.140250509

- Xie, Jiangbing; Hsieh, You-Lo (25 July 2003). "Thermosensitive poly(n-isopropylacrylamide) hydrogels bonded on cellulose supports". Journal of Applied Polymer Science. 89 (4): 999–1006. doi:10.1002/app.12206

- Polymer-Architecture, chemical-engineering, technical-references: idconline.com, Retrieved 22 May 2018

- editors, Susheel Kalia, M.W. Sabaa, (2013). Polysaccharide based graft copolymers (1., 2013 ed.). Heidelberg: Springer. ISBN 9783642365652

# Permissions

All chapters in this book are published with permission under the Creative Commons Attribution Share Alike License or equivalent. Every chapter published in this book has been scrutinized by our experts. Their significance has been extensively debated. The topics covered herein carry significant information for a comprehensive understanding. They may even be implemented as practical applications or may be referred to as a beginning point for further studies.

We would like to thank the editorial team for lending their expertise to make the book truly unique. They have played a crucial role in the development of this book. Without their invaluable contributions this book wouldn't have been possible. They have made vital efforts to compile up to date information on the varied aspects of this subject to make this book a valuable addition to the collection of many professionals and students.

This book was conceptualized with the vision of imparting up-to-date and integrated information in this field. To ensure the same, a matchless editorial board was set up. Every individual on the board went through rigorous rounds of assessment to prove their worth. After which they invested a large part of their time researching and compiling the most relevant data for our readers.

The editorial board has been involved in producing this book since its inception. They have spent rigorous hours researching and exploring the diverse topics which have resulted in the successful publishing of this book. They have passed on their knowledge of decades through this book. To expedite this challenging task, the publisher supported the team at every step. A small team of assistant editors was also appointed to further simplify the editing procedure and attain best results for the readers.

Apart from the editorial board, the designing team has also invested a significant amount of their time in understanding the subject and creating the most relevant covers. They scrutinized every image to scout for the most suitable representation of the subject and create an appropriate cover for the book.

The publishing team has been an ardent support to the editorial, designing and production team. Their endless efforts to recruit the best for this project, has resulted in the accomplishment of this book. They are a veteran in the field of academics and their pool of knowledge is as vast as their experience in printing. Their expertise and guidance has proved useful at every step. Their uncompromising quality standards have made this book an exceptional effort. Their encouragement from time to time has been an inspiration for everyone.

The publisher and the editorial board hope that this book will prove to be a valuable piece of knowledge for students, practitioners and scholars across the globe.

# Index

www.ingramcontent.com/pod-product-compliance
Lightning Source LLC
Chambersburg PA
CBHW061954190326
41458CB00009B/2870